T0320360

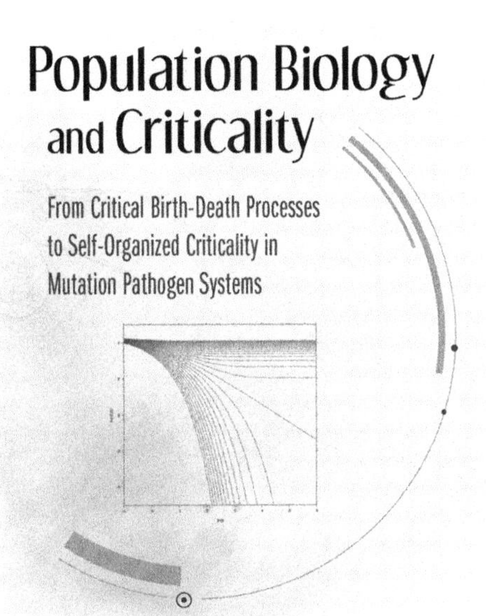
Population Biology
and Criticality
From Critical Birth-Death Processes
to Self-Organized Criticality in
Mutation Pathogen Systems

Population Biology and Criticality

From Critical Birth-Death Processes to Self-Organized Criticality in Mutation Pathogen Systems

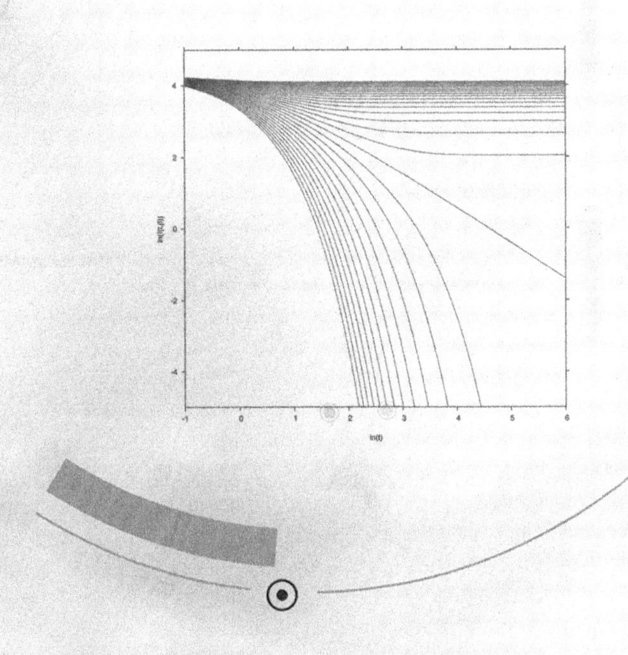

Nico Stollenwerk

Universidade de Lisboa, Portugal

Vincent Jansen

Royal Holloway, University of London

Imperial College Press

ICP

Published by

Imperial College Press
57 Shelton Street
Covent Garden
London WC2H 9HE

Distributed by

World Scientific Publishing Co. Pte. Ltd.
5 Toh Tuck Link, Singapore 596224
USA office: 27 Warren Street, Suite 401-402, Hackensack, NJ 07601
UK office: 57 Shelton Street, Covent Garden, London WC2H 9HE

British Library Cataloguing-in-Publication Data
A catalogue record for this book is available from the British Library.

POPULATION BIOLOGY AND CRITICALITY
From Critical Birth-Death Processes to Self-Organized Criticality in Mutation Pathogen Systems

ISBN-13 978-1-84816-401-7
ISBN-10 1-84816-401-7

Printed in Singapore.

Preface

This book describes novel theories of mutation pathogen systems showing critical fluctuations as a paradigmatic example of an application of the mathematics of critical phenomena to the life sciences. It will enable the reader to understand the implications and future impact of these findings, and at same time allow them to actively follow the mathematical tools and scientific origins of critical phenomena. Hopefully this will lead to further fruitful applications of the mathematics of critical phenomena in other fields of the life sciences.

We show through explicit calculations the basis for understanding critical fluctuations as a genuine effect of stochastic dynamic systems, including the original background in statistical physics. Birth and death processes occur in many applications in the life sciences, in ecological and epidemiological systems and cancer growth, for example. In such simple systems the critical threshold appears and can be analyzed in all its different aspects.

Guided by our own research experience we then proceed to a system of pathogens evolving towards a critical state in a multi-mutant system. Although we concentrate on a special case study, the mechanism will certainly have much wider implications theoretically, understanding the mechanisms leading to systems with large fluctuations, as well as empirically, as in many real world systems such large fluctuations are observed but their mechanism often not recognized as such or not well understood.

To enable researchers in the various fields of application to analyze their systems appropriately we show explicitly the calculations and mathematical tools involved, which are now widespread in the textbook literature and the original scientific literature in journals. Since they are often spread over a wide range of scientific fields they use very different language and notations.

From our own teaching experience we see the necessity of taking the reader from basic notions of stochastic dynamics, often taught to physics students and probability theoreticians but rarely to students from the life sciences, to a detailed analysis of the birth and death process, which is often taught to biologists and other life scientists, but rarely in the aspect of criticality. The field of critical phenomena in birth and death processes itself is a branch of rather high specialization in statistical physics with loose applications mainly in condensed matter physics or no application directly in mind.

However, when looking at real world data, large fluctuations are often observed, and cannot always be traced back to large external influences, hence the need to search for possible internal mechanisms. These internal mechanisms are given by the theory of critical phenomena, as we will explicitly demonstrate in our paradigmatic case study.

This book contains two parts. The first includes an introduction to stochastic dynamic systems and criticality, the second explores the application of mutation pathogen systems as a case study for wider applications.

We thank our students, José Martins, Maíra Aguiar, Rui Gonçalvez and Sander van Noort, for many discussions and calculations, and many collegues of ours for inspiration, especially but by no means exclusively Friedhelm Drepper, Minus van Baalen, Martin Maiden, Alberto Pinto, Gabriela Gomes, Frank Hilker, Lewi Stone, Walter Nadler and Luis Sanchez.

N. Stollenwerk and V. Jansen
Lisbon and London, August 2009

Contents

Chapter 1

From Deterministic to Stochastic Dynamics

Since we describe genuinely stochastic effects, as critical fluctuations are, we start by introducing the basic notions to describe any stochastic dynamics, working out the Perron–Frobenius equation as the simplest stochastic dynamic, a time-discrete Markov process. Two examples will be used. We start with a deterministic map, the Ulam map, to see the notions of probability in action to describe a dynamic process. We then generalize to the simplest truly stochastic process, the autoregressive process of order one, the AR(1)-process, i.e. only one time step in the past determines the next step.

For this AR(1)-process, essentially all the terminology of probabilistic dynamics can be worked out, including parameter estimation in maximum likelihood form and in the Bayesian statistics.

From the Perron–Frobenius equation we can easily generalize to birth–death processes in general, the birth and death rates being transition probabilities per time. We can then describe time-discrete processes but, more interestingly, also time-continuous stochastic processes in the form of a master equation, first non-spatially for the total number of individuals, then in Chapter 2 also spatially, introducing variables for each lattice point for having just one or no individual present.

1.1 Basic Probability Theory: The Tool Box

For the description of any basic stochastic dynamic process we will need the following essential notions from probability theory: the joint distribution, the marginal and the conditional distribution. More detailed descriptions can be found in the literature, so we only need the practical rules given here.

A joint distribution $p(x, y)$ is the probability of finding values for a variable x and a variable y. The probability density is normalized in the sense that summing up over x and over y gives unity, hence $\int \int p(x, y)\, dx\, dy = 1$. As a starting point for probabilistic considerations, in the next section we will use the joint distribution of two or several variables

$$p(x, y) \quad . \tag{1.1}$$

The marginal distribution $p(x) = \int p(x, y)\, dy$ gives the probability for the variable x, independently of what the value for y might be. Hence, the first equation, which we will often use in the following, is

$$p(x) = \int p(x, y)\, dy \quad . \tag{1.2}$$

The conditional distribution $p(x|y)$ is the probability of x, given that the variable y takes a certain specified value. It is linked with the joint distribution through Bayes' rule, which is $p(x, y) = p(x|y) \cdot p(y)$. Sometimes this equation is given as the definition of the conditional distribution. However, we prefer the above operational definition so that Bayes' rule is a result. Hence, as the second often used equation we have

$$p(x, y) = p(x|y) \cdot p(y) \quad . \tag{1.3}$$

The distribution that an event x_0 is given with certainty is $p(x) = \delta(x - x_0)$, where the delta-function $\delta(x - x_0)$ is used in the operational equation $\int_a^b f(x) \cdot \delta(x - x_0)\, dx = f(x_0)$ for any function $f(x)$ as long as x_0 is in the interval between a and b, otherwise the integral is zero. However mathematically rigorous the delta-function might be, the use under the integral we apply here is well defined and standard in physics. The following sections will demonstrate the success of this notion, as well as the other rules shown here. Hence, as the third often used equation we have the integration over a delta-function

$$\int_a^b f(x) \cdot \delta(x - x_0)\, dx = f(x_0) \quad . \tag{1.4}$$

In the next section, we will give as a first application the description of the Ulam map; a purely deterministic map which, however, shows its full character only when described in terms of a stochastic system [Ulam and von Neumann (1947)].

This example is interesting for several reasons. Firstly, the above mentioned tools Eq. (1.1) to Eq. (1.4) are all we need to describe the Ulam map in its stochastic behavior, and show the close analogy to a purely deterministic description. Secondly, the Ulam map can easily be iterated on a computer or even on a programmable pocket calculator, without using any random generator, and still gives a smooth random distribution. Thirdly, since the Ulam map is one of the family of quadratic maps studied extensively in the theory of deterministic chaos, nowadays often described as the simplest equation with rich dynamics behavior, most readers will be already familiar to some extent with the basic concepts of the quadratic map, the more since it has been introduced very early to biology [May (1974)] and was extensively treated by [Feigenbaum (1978)]. For a recent review of the richness in qualitative aspects of the quadratic map see [Beck and Schlögl (1993)].

1.2 Stochastic Description of a Deterministic System: The Ulam Map

The Ulam map is given by

$$x_{n+1} = 4x_n(1 - x_n) =: f(x_n) \tag{1.5}$$

as a deterministic map, mapping a starting value x_0 into x_1 into x_2 etc. [Ulam and von Neumann (1947)]. The map exhibits deterministic chaotic motion on the interval $[0, 1]$.

The stochastic version, mapping an ensemble $p(x_0)$ of starting values x_0 into an ensemble $p(x_1)$ for the following value x_1 etc., is given through the basic stochastic notions of marginal distribution $p(x_{n+1})$ from a joint distribution $p(x_{n+1}, x_n)$ and Bayes' rule, using the delta-function for the conditional distribution $p(x_{n+1}|x_n) = \delta(x_{n+1} - f(x_n))$ via

$$p(x_{n+1}) = \int_0^1 p(x_{n+1}, x_n)\, dx_n \tag{1.6}$$

from the definition of marginal distributions and with Bayes' rule

$$p(x_{n+1}, x_n) = p(x_{n+1}|x_n) \cdot p(x_n) \quad . \tag{1.7}$$

Now, since the deterministic equation for the Ulam map is given by Eq. (1.5), we use the conditional distribution $p(x_{n+1}|x_n) = \delta(x_{n+1} - f(x_n))$ to specify

$$p(x_{n+1}|x_n) = \delta(x_{n+1} - (4x_n(1 - x_n))) \tag{1.8}$$

obtaining the dynamic equation for the stochastic Ulam process

$$p(x_{n+1}) = \int_0^1 \delta(x_{n+1} - (4x_n(1-x_n))) \cdot p(x_n)\, dx_n \qquad (1.9)$$

which maps a given $p(x_n)$ into $p(x_{n+1})$.

For simplicity we start with a uniform distribution for the starting value x_0, hence

$$p(x_0) = 1 \qquad \text{for} \quad x_0 \in [0,1] \qquad (1.10)$$

and calculate $p(x_1)$ via the dynamics Eq. (1.9)

$$p(x_1) = \int_0^1 \delta(x_1 - \underbrace{f(x_0)}_{=:y}) \cdot \underbrace{p(x_0)}_{=1}\, dx_0 \qquad (1.11)$$

substituting $f(x_0)$ by $y := 4x_0(1-x_0)$, hence $dy/dx_0 = 4 - 8x_0$. From the equation $y := 4x_0(1-x_0)$ we obtain $x_0(y)$ such that

$$\begin{aligned} x_0 &= \tfrac{1}{2}(1 - \sqrt{1-y}) \quad \text{for } x_0 \in \left[0, \tfrac{1}{2}\right] \\ x_0 &= \tfrac{1}{2}(1 + \sqrt{1-y}) \quad \text{for } x_0 \in \left[\tfrac{1}{2}, 1\right] \end{aligned} \qquad (1.12)$$

giving

$$dx_0 = \frac{1}{\left(\frac{dy}{dx_0}\right)}\, dy = \frac{1}{4 - 8x_0(y)}\, dy = \frac{dy}{\pm 4\sqrt{1-y}} \quad . \qquad (1.13)$$

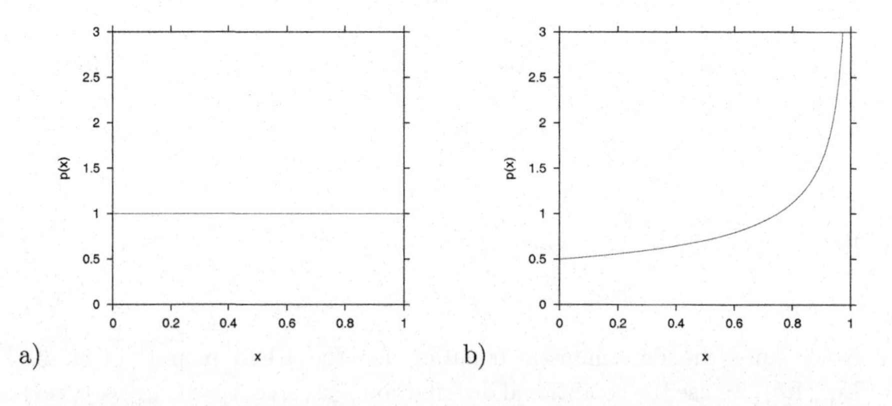

Fig. 1.1 a) The starting distribution of the Ulam map can be a uniform distribution on the interval between zero and one. It is mapped in the first iteration into the distribution given in b).

So it is

$$\begin{aligned} p_1(x_1) &= \int\limits_{y(0)}^{y(\frac{1}{2})} \delta(x_1 - y) \cdot \frac{dy}{4\sqrt{1-y}} + \int\limits_{y(\frac{1}{2})}^{y(1)} \delta(x_1 - y) \cdot \frac{dy}{-4\sqrt{1-y}} \\ &= \int\limits_0^1 \delta(x_1 - y) \cdot \frac{dy}{4\sqrt{1-y}} + \int\limits_1^0 \delta(x_1 - y) \cdot \frac{dy}{-4\sqrt{1-y}} \\ &= \frac{2}{4\sqrt{1-x_1}} \end{aligned} \tag{1.14}$$

using $\int\limits_1^0 f(x)dx = -\int\limits_0^1 f(x)dx$ for any function $f(x)$ and symmetry of the delta-function $\delta(x-y) = \delta(y-x)$. Hence, we obtain

$$p_1(x_1) = \frac{2}{4\sqrt{1-x_1}} \tag{1.15}$$

as shown in Fig. 1.1b), whereas 1.1a) shows the initial uniform distribution $p_0(x_0) = 1$.

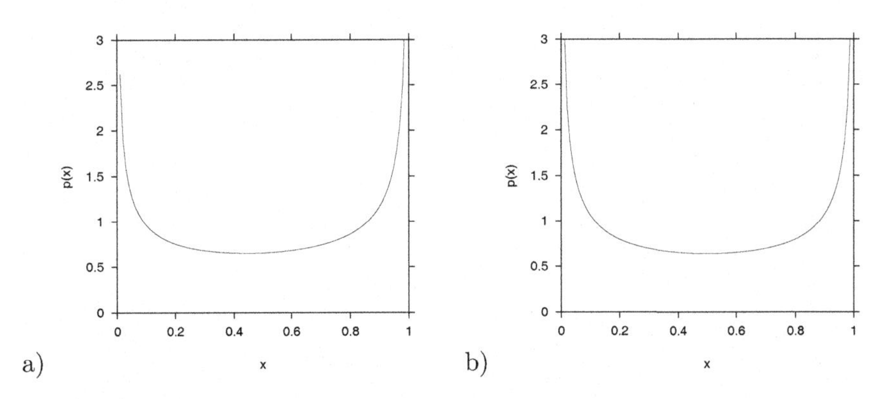

Fig. 1.2 After two iterations of the initial uniform distribution under the Ulam map, the resulting distribution a) shows already the typical U-shape, which has become famous for the stationary distribution of the Ulam map. The stationary distribution $p^*(x)$ is shown in b).

Similarly, we can calculate $p_2(x_2)$ as

$$p_2(x_2) = \frac{1}{4\sqrt{2} \cdot \sqrt{1-x_2}} \cdot \left(\frac{1}{\sqrt{1+\sqrt{1-x_2}}} + \frac{1}{\sqrt{1-\sqrt{1-x_2}}} \right) \quad . \tag{1.16}$$

In Fig. 1.2a) the distribution $p_2(x_2)$ is shown. In Fig. 1.2b) we show after $n \to \infty$ the stationary distribution

$$p^*(x_n) = \frac{1}{\pi\sqrt{x_n(1-x_n)}} \quad . \tag{1.17}$$

obeying the fixed point equation

$$p^*(x_{n+1}) = \int_0^1 \delta\left(x_{n+1} - 4x_n(1-x_n)\right) \cdot p^*(x_n)\, dx_n \tag{1.18}$$

for the same function p^* in both sides of the Perron–Frobenius equation. The calculations for the stationary distribution are shown in more detail in Appendix A, where its chaotic nature also can be demonstrated by calculating the Lyapunov exponent, using the stationary distribution of the Ulam map.

1.3 A Fully Stochastic Dynamic System: The AR(1)-Process

As a simple example of a truly stochastic dynamic system, we first consider the linear autoregressive order one model, AR(1), which is also called the time-discrete Ornstein–Uhlenbeck process. This is the dynamical process

$$x_{n+1} = a\, x_n + b + \sigma\, \varepsilon_n \tag{1.19}$$

for times $n \in \mathbb{N}$ and state $x_n \in \mathbb{R}$ with parameters a, b and σ and with independent Gauß normal noise ε_n, i.e. with probability density

$$p(\varepsilon_n) = \frac{1}{\sqrt{2\pi}} e^{-\frac{1}{2}(\varepsilon_n)^2} \tag{1.20}$$

and stochastic independence, i.e.

$$p(\varepsilon_n, \varepsilon_{n+k}) = p(\varepsilon_n) \cdot p(\varepsilon_{n+k})$$

for all times n and k. It is a, $b \in \mathbb{R}$, $\sigma \in \mathbb{R}_+$, $\varepsilon_n \in \mathbb{R}$.

From first principles the marginal distribution $p(x_{n+1})$ is determined from the joint distribution $p(x_{n+1}, x_n, \varepsilon_n)$ via the integration

$$p(x_{n+1}) = \int_{-\infty}^{\infty}\int_{-\infty}^{\infty} p(x_{n+1}, x_n, \varepsilon_n)\, dx_n\, d\varepsilon_n$$

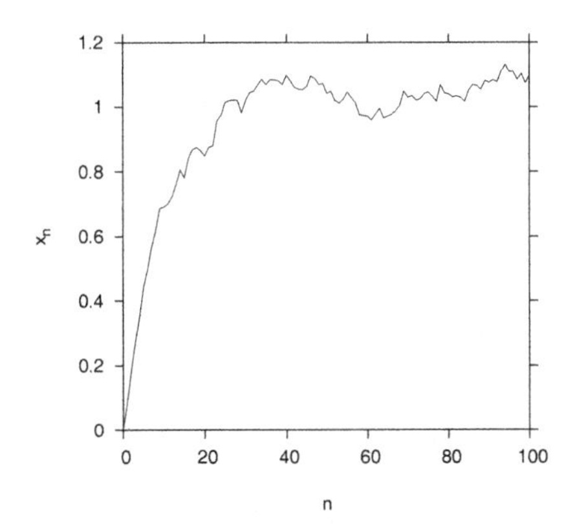

Fig. 1.3 From an AR(1)-model with parameters $a = 0.9$ and $b = 0.1$, $\sigma = 0.02$ a single trajectory is simulated. The transient behavior into the stationary behavior is clearly visible. One hundered data points are generated ($N = 100$).

and with Bayes' rule (see in [van Kampen (1992)], on p. 10) the joint probability is given as a product with conditional probabilities by

$$p(x_{n+1}, x_n, \varepsilon_n) = p(x_{n+1}|x_n, \varepsilon_n) \cdot p(x_n, \varepsilon_n) \quad .$$

The conditional probability, which is here a transition probability of the dynamics (1.19), is given by the Dirac delta-function

$$p(x_{n+1}|x_n, \varepsilon_n) = \delta(x_{n+1} - f(x_n, \varepsilon_n)) \quad .$$

Furthermore, the noise ε_n is independent of the state x_n, which means

$$p(x_n, \varepsilon_n) = p(x_n) \cdot p(\varepsilon_n) \quad .$$

From this, we obtain the evolution equation of the dynamics as given by $x_{n+1} = f(x_n, \varepsilon_n)$, hence the Perron–Frobenius equation for the distribution $p_n(x_n)$ is

$$p_{n+1}(x_{n+1}) = \int_{-\infty}^{\infty} \int_{-\infty}^{\infty} \delta(x_{n+1} - f(x_n, \varepsilon_n)) \cdot p_n(x_n) \cdot p_\varepsilon(\varepsilon_n) \, dx_n \, d\varepsilon_n \quad .$$

Executing the integration over ε_n, using the features of the delta-function as described above, gives

$$p_{n+1}(x_{n+1}) = \int_{-\infty}^{\infty} \frac{1}{\sigma\sqrt{2\pi}} e^{-\frac{1}{2\sigma^2}(x_{n+1} - (ax_n + b))^2} \cdot p_n(x_n) \, dx_n \quad . \tag{1.21}$$

Hence the transition probability

$$p(x_{n+1}|x_n) = \frac{1}{\sigma\sqrt{2\pi}} e^{-\frac{1}{2\sigma^2}(x_{n+1} - (ax_n + b))^2} \tag{1.22}$$

has now a Gaussian form, and is no longer a delta-function, as it was in the case of the Ulam–map, a deterministic system.

In summary, we obtain the same form of the Perron–Frobenius equation as in the previous deterministic system by integrating over the noise variable

$$\begin{aligned} p_{n+1}(x_{n+1}) &= \int \underbrace{\left(\int p(x_{n+1}|x_n, \varepsilon_n) p_\varepsilon(\varepsilon_n) \, d\varepsilon_n \right)}_{=p(x_{n+1}|x_n)} p_t(x_n) \, dx_n \\ &= \int p(x_{n+1}|x_n) p_n(x_n) \, dx_n \quad . \end{aligned} \tag{1.23}$$

From here, different general aspects can be studied in more detail. Firstly, the stationary distribution of Eq. (1.21) can be calculated explicitly. Secondly, the joint distribution of many "data points" can be constructed, leading to parameter estimation in the form of likelihood maximization or Bayesian estimation using, for example, conjugate priors (see Appendix B).

1.4 From Perron–Frobenius to Master Equation

The Perron–Frobenius equation was previously derived by integrating over the noise variable, see Eq. (1.23),

$$\begin{aligned} p_{t+\Delta t}(x_{t+\Delta t}) &= \int \underbrace{\left(\int p(x_{t+\Delta t}|x_t, \varepsilon_t) p_\varepsilon(\varepsilon_t) \, d\varepsilon_t \right)}_{=p(x_{t+\Delta t}|x_t)} p_t(x_t) \, dx_t \\ &= \int p(x_{t+\Delta t}|x_t) p_t(x_t) \, dx_t \quad . \end{aligned} \tag{1.24}$$

We have just changed the iteration variable n to a time variable t and iteration step $n + 1$ to $t + \Delta t$. From this form we can easily derive a time-continuous version for $\Delta t \to 0$ in the following way

$$\begin{aligned} \tfrac{d}{dt} p(x_{t+\Delta t}) &\approx \tfrac{1}{\Delta t}(p_{t+\Delta t}(x_{t+\Delta t}) - p_t(x_{t+\Delta t})) \\ &= \tfrac{1}{\Delta t}\left(\int p(x_{t+\Delta t}|x_t) p_t(x_t) \, dx_t - p_t(x_{t+\Delta t}) \right) \quad . \end{aligned} \tag{1.25}$$

While the index t in the probability function p_t labels a change in time from one functional form to another $p_{t+\Delta t}$, the index t or $t+\Delta t$ is simply a label for the integration or at which point of the function p we look at. We can also replace $x_{t+\Delta t}$ by x only, and also x_t by another variable $\tilde{x}$. Thus we have

$$\begin{aligned} \tfrac{d}{dt}p(x) &\approx \tfrac{1}{\Delta t}(p_{t+\Delta t}(x) - p_t(x)) \\ &= \tfrac{1}{\Delta t}\left(\int p(x|\tilde{x})p_t(\tilde{x})\, d\tilde{x} - p_t(x)\right) \quad . \end{aligned} \tag{1.26}$$

Now, the the second part of the final expression we can consider to be a marginal distribution again from $p(\tilde{x}, x)$. Applying Bayes' rule once more we get $p(\tilde{x}, x) = p(\tilde{x}|x) \cdot p(x)$, hence

$$p_t(x) = \int p(\tilde{x}|x)p_t(x)\, d\tilde{x} \quad . \tag{1.27}$$

We finally obtain the master equation

$$\begin{aligned} \frac{d}{dt}p(x) = &\int \left(\frac{1}{\Delta t}p(x|\tilde{x})\right) p_t(\tilde{x})\, d\tilde{x} \\ &- \int \left(\frac{1}{\Delta t}p(\tilde{x}|x)\right)\, d\tilde{x} \cdot p_t(x) \quad . \end{aligned} \tag{1.28}$$

With the transition rates $w_{x,\tilde{x}} := \frac{1}{\Delta t}p(x|\tilde{x})$ we obtain

$$\frac{d}{dt}p(x) = \int w_{x,\tilde{x}} \cdot p_t(\tilde{x})\, d\tilde{x} - \int w_{\tilde{x},x} \cdot p_t(x)\, d\tilde{x} \tag{1.29}$$

the master equation for continuous state variable x, a time-continuous Markov process.

1.5 A First Example of a Master Equation: The Linear Infection Model

We have derived the master equation for a continuous state variable as in Eq. (1.29), an integro-differential equation, in order to be able to easily continue to derive the Fokker–Planck equation from it, a partial differential equation, which again can be treated rigorously to some extent.

Here we will describe a first application to population dynamics or to epidemiology, for which we need the state-discrete version of the master equation as the starting point

$$\frac{dp(\underline{n},t)}{dt} = \sum_{\underline{\tilde{n}} \neq \underline{n}} w_{\underline{n},\underline{\tilde{n}}} \; p(\underline{\tilde{n}},t) - \sum_{\underline{\tilde{n}} \neq \underline{n}} w_{\underline{\tilde{n}},\underline{n}} \; p(\underline{n},t) \quad . \tag{1.30}$$

The state variable $\underline{n}$, previously a continuous variable x, now only has discrete values.

To specify an example, let us look at N individuals, which are initially susceptible, $S = N$. The susceptibles can become infected independently of each other, i.e. the infection occurs from outside and not from direct contacts. The number of infected is given by I and at all times we have $S + I = N$. The state vector could be written as $\underline{n} = (S, I)$, but since fixing I we know S for sure, just one variable is sufficient.

Hence the master equation becomes

$$\frac{dp(I,t)}{dt} = \sum_{\tilde{I} \neq I} w_{I,\tilde{I}} \; p(\tilde{I}) - \sum_{\tilde{I} \neq I} w_{\tilde{I},I} \; p(I) \tag{1.31}$$

with transition rates still to be specified. It is for $\underline{n} = (S, I)$

$$\begin{aligned} w_{(S-1,I+1),(S,I)} &= \beta \cdot S \\ w_{(S,I,),(S+1,I-1)} &= \beta \cdot (S+1) \end{aligned}$$

or just for the variable I

$$\frac{dp(I,t)}{dt} = \beta(N-(I-1)) \; \cdot \; p(I-1,t) - \beta(N-I) \; \cdot \; p(I,t) \quad . \tag{1.32}$$

As boundary equations we have for $I = 0$

$$\frac{dp(I=0,t)}{dt} = -\beta N \; \cdot \; p(I=0,t) \tag{1.33}$$

and for $I = N$

$$\frac{dp(I=N,t)}{dt} = \beta \; \cdot \; p(I=N-1,t) \quad . \tag{1.34}$$

In the limit for large populations $N \to \infty$ and small times or few infected, hence $\beta(N-(I-1)) \approx \beta(N-I) \approx \beta N =: \lambda$ the master equation reduces further to

$$\frac{dp(I,t)}{dt} = \lambda \; \cdot \; p(I-1,t) - \lambda \; \cdot \; p(I,t) \quad . \tag{1.35}$$

1.5.1 *Solving the first example of a master equation*

We solve

$$\frac{dp(I,t)}{dt} = \lambda \cdot p(I-1,t) - \lambda \cdot p(I,t) \quad . \tag{1.36}$$

for $I = 0, 1, 2, ...\infty$ and constant transition rate λ explicitly. Initial conditions are given, e.g. by $p(I = 0, t_0 = 0) = 1$, $p(I \neq 0, t_0 = 0) = 0$; hence a sharp initial distribution with only mass at $I = 0$.

Looking at the equation for $I = 0$, and labelling $p(I = 0, t) =: p_0(t)$ we find that

$$\frac{dp_0(t)}{dt} = -\lambda \cdot p_0(t) \quad . \tag{1.37}$$

This is just a linear ordinary differential equation (ODE), hence easy to solve. It is

$$\frac{1}{p_0}\, dp_0 = -\lambda\, dt \tag{1.38}$$

and integrated from t_0 to t_1

$$\int_{p_0(t_0)}^{p_0(t_1)} \frac{1}{p_0}\, dp_0 = -\lambda \int_{t_0}^{t_1} dt \tag{1.39}$$

hence

$$\Big[\ln p_0 \Big]_{p_0(t_0)}^{p_0(t_1)} = -\lambda \Big[t \Big]_{t_0}^{t_1} \tag{1.40}$$

or

$$\ln p_0(t_1) - \ln p_0(t_0) = -\lambda(t_1 - t_0) \quad . \tag{1.41}$$

Thus, as general solution for $p_0(t)$, dropping the index at the upper integration limit t_1, we obtain

$$p_0(t) = p_0(t_0) \cdot e^{-\lambda(t-t_0)} \quad . \tag{1.42}$$

Taking the initial conditions $t_0 = 0$ and $p_0(t_0) = 1$ into account, we simply get

$$p_0(t) = e^{-\lambda t} \quad . \tag{1.43}$$

Next, we look at the equation for $I = 1$. Labeling $p(I = 1, t) =: p_1(t)$ we find

$$\frac{dp_1}{dt} = -\lambda \cdot p_1 + \lambda \cdot p_0(t) \tag{1.44}$$

with the already known solution $p_0(t) = p_0(t_0) \cdot e^{-\lambda(t-t_0)}$. Thus we have an inhomogeneous linear ODE, which can be solved by the method of varying constant. The solution of the linear homogeneous ODE $dp_1/dt = -\lambda p_1$ is, as just seen, $p_1(t) = C \cdot e^{-\lambda t}$ with an integration constant C. Hence the solution for the inhomogeneous linear ODE $dp_1/dt = -\lambda p_1 + f(t)$ for any time-dependent function $f(t)$ is $p_1(t) = C(t) \cdot e^{-\lambda t}$ where the previous constant C is now a time-dependent function $C(t)$ to be determined, hence the term "variation of constant".

In detail,

$$\frac{d}{dt}p_1 = \left(\frac{d}{dt}C(t)\right) \cdot e^{-\lambda t} + C(t) \cdot \left(\frac{d}{dt}e^{-\lambda t}\right) \tag{1.45}$$

gives

$$\frac{d}{dt}p_1 = -\lambda C e^{-\lambda t} + \left(\frac{d}{dt}C(t)\right) \cdot e^{-\lambda t} \tag{1.46}$$

Comparing this with the original ODE, Eq. (1.44), shows that

$$\left(\frac{d}{dt}C(t)\right) \cdot e^{-\lambda t} = f(t) \tag{1.47}$$

in general, and here $f(t)$ has the form

$$f(t) = \lambda p_0(t_0) e^{-\lambda(t-t_0)} \tag{1.48}$$

hence, we have to solve the following ODE

$$\frac{d}{dt}C(t) = e^{\lambda t} f(t) = \lambda p_0(t_0) e^{\lambda t_0} \quad . \tag{1.49}$$

Again we integrate

$$\int_{C_0(t_0)}^{C_0(t_1)} dC_0 = \lambda p_0(t_0) e^{\lambda t_0} \cdot \int_{t_0}^{t_1} dt \tag{1.50}$$

and omitting the index of t_1, the upper integration boundary, we obtain

$$C(t) = C(t_0) + \lambda p_0(t_0) e^{\lambda t_0} \cdot (t - t_0) \quad . \tag{1.51}$$

With the initial condition $p(I = 1, t_0) = p_1(t_0)$ we have now the ansatz as $p_1(t_0) = C(t_0) \cdot e^{-\lambda t_0}$, hence

$$C(t_0) = e^{\lambda t_0} \cdot p_1(t_0) \quad , \tag{1.52}$$

which for $p_1(t) = C(t)e^{-\lambda t}$ finally gives the result

$$p_1(t) = p_1(t_0) \cdot e^{-\lambda(t-t_0)} + p_0(t_0) \cdot \lambda \cdot (t - t_0) \cdot e^{-\lambda(t-t_0)} \quad , \tag{1.53}$$

or taking all initial conditions, $t_0 = 0$, $p_0(t_0) = 1$ and $p_1(t_0) = 0$ into account, we simply get

$$p_1(t) = (\lambda t) \cdot e^{-\lambda t} \quad . \tag{1.54}$$

From Eq. (1.50) and Eq. (1.51) we can easily generalize to a recursion formula for $p_I(t)$ from $p_{I-1}(t)$ which is

$$p_I(t) = p_I(t_0)e^{-\lambda(t-t_0)} + \lambda e^{-\lambda t} \int_{t_0}^{t_1} e^{\lambda\tau} p_{I-1}(\tau)\, d\tau \quad . \tag{1.55}$$

Then, for the special initial conditions we obtain as final result

$$p(I, t) = \frac{(\lambda t)^I}{I!}\, e^{-\lambda t} \tag{1.56}$$

which is a Poisson distribution. Hence the process defined by Eq. (1.36) is also known as the Poisson process and is widely studied in natural sciences [Bharucha-Reid (1960)].

1.5.2 *Solution of the linear infection model*

With the methods described in Section 1.5.1 for the Poisson process, one can also solve the original linear infection model Eq. (1.32) to obtain

$$p(I, t) = \binom{N}{I} \left(e^{-\beta t}\right)^{N-I} \left(1 - e^{-\beta t}\right)^I \tag{1.57}$$

which is a binomial distribution for N trials with I successes and success probability $\left(1 - e^{-\beta t}\right)$. In technical terms, this process is also known as the linear death process, refering to the death of the initial N susceptibles [Bharucha-Reid (1960)].

The master equation is a set of ordinary differential equations linear in probabilities p. From $p(I = 0, t)$ ... $p(I = N, t)$ we construct a vector $\underline{p}$.

From the transition rates $w_{I,\tilde{I}}$ we construct a matrix A, such that the full system is given by the equation

$$\frac{d}{dt}\underline{p} = A \cdot \underline{p} \quad . \tag{1.58}$$

It has as its formal solution

$$\underline{p}(t) = e^{At} \cdot \underline{p}(t_0) \tag{1.59}$$

with the matrix exponential $e^{At} := \sum_{\nu=1}^{\infty} \frac{t^\nu}{\nu!} A^\nu$ where A^ν is just the product of square matrices, subsequently applied to the vector of initial conditions $\underline{p}(t_0)$.

From this solution, a likelihood function can be constructed and parameters estimated (see Appendix C for an example of a non-linear master equation, i.e. non-linear in the state variable I and not in p).

1.5.3 *Mean value and its dynamics*

Now we look at the mean value of the linear infection model

$$\langle I \rangle := \sum_{I=0}^{N} I \cdot p(I,t) \quad . \tag{1.60}$$

Inserting the solution of the distribution $p(I,t)$ for the linear infection model we obtain

$$\begin{aligned}
\langle I \rangle &= \sum_{I=0}^{N} I \binom{N}{I} \left(e^{-\beta t}\right)^{N-I} \left(1 - e^{-\beta t}\right)^{I} \\
&= \sum_{I=0}^{N} I \, \frac{N!}{I!(N-I)!} \left(e^{-\beta t}\right)^{N-I} \left(1 - e^{-\beta t}\right)^{I} \\
&= N \left(1 - e^{-\beta t}\right) \\
&\quad \cdot \sum_{I=0}^{N} \frac{(N-1)!}{(I-1)!(N-1-(I-1))!} \left(e^{-\beta t}\right)^{N-1-(I-1)} \left(1 - e^{-\beta t}\right)^{I-1}
\end{aligned} \tag{1.61}$$

and with the binomial formula $(a+b)^n = \sum_{k=0}^{n} \binom{n}{k} a^k \cdot b^{n-k}$ we obtain with the substitutions $k := I-1$ and $n := N-1$

$$\begin{aligned} \langle I \rangle &= N\left(1-e^{-\beta t}\right) \sum_{k=0}^{n} \binom{n}{k} \left(e^{-\beta t}\right)^{n-k} \left(1-e^{-\beta t}\right)^k \\ &= N\left(1-e^{-\beta t}\right) \end{aligned} \tag{1.62}$$

hence

$$\langle I \rangle = N\left(1-e^{-\beta t}\right) \quad . \tag{1.63}$$

1.5.4 *Mean dynamics*

From the definition of the mean value

$$\langle I \rangle := \sum_{I=0}^{N} I\, p(I,t) \tag{1.64}$$

we can also derive a dynamic equation for the mean

$$\frac{d}{dt}\langle I \rangle = \sum_{I=0}^{N} I\, \frac{d}{dt} p(I,t) \quad . \tag{1.65}$$

Now we can insert for $\frac{d}{dt}p(I,t)$ the full master equation (and keeping the boundary equations in mind)

$$\frac{dp(I,t)}{dt} = \beta(N-(I-1)) \cdot p(I-1,t) - \beta(N-I) \cdot p(I,t) \tag{1.66}$$

and obtain

$$\begin{aligned} \frac{d}{dt}\langle I \rangle &= \sum_{I=0}^{N} I\, \frac{d}{dt} p(I,t) \\ &= \sum_{I=1}^{N} I\, \beta(N-(I-1)) \cdot p(I-1,t) - \sum_{I=0}^{N} I\, \beta(N-I) \cdot p(I,t) \quad . \end{aligned} \tag{1.67}$$

Substituting in the summation $\tilde{I} := I - 1$ or $I = \tilde{I} + 1$, and taking care of the summation boundaries, we obtain

$$\begin{aligned}\frac{d}{dt}\langle I\rangle &= \beta\sum_{\tilde{I}=0}^{N-1}(\tilde{I}+1)(N-\tilde{I})\,p(\tilde{I},t) - \beta\sum_{I=0}^{N} I\,(N-I)\;\cdot\;p(I,t)\\ &= \beta\cdot N\sum_{\tilde{I}=0}^{N} p(\tilde{I},t) - \beta\sum_{I=0}^{N} I\;\cdot\;p(I,t)\\ &= \beta\cdot(N-\langle I\rangle)\end{aligned} \tag{1.68}$$

and hence

$$\frac{d}{dt}\langle I\rangle = \beta\cdot(N-\langle I\rangle) \tag{1.69}$$

as a closed system of an ODE only containing the mean value $\langle I\rangle$ itself.

This ODE, Eq. (1.69), we can integrate like any other ODE with the techniques seen in Section 1.5.1 and directly obtain

$$\langle I\rangle = N\left(1 - e^{-\beta t}\right) \tag{1.70}$$

without solving the stochastic master equation explicitly. Compare the solution with the previously obtained solution Eq. (1.63).

While for a linear stochastic system, that is linear in state variable I, we can obtain a closed ODE with the insertion of the master equation, this is not the case for non-linear processes such as those investigated extensively in this book. All the complexity possible in such non-linear systems is hidden in fine detail, e.g. a quadratic term in I, whereas the stochastic system is still a linear ODE system in the probability function $p(I)$.

1.6 The Birth and Death Process, a Non-Linear Stochastic System

The birth–death process, here for individuals I and empty spaces S (or in epidemiology for infected I and susceptibles S)

$$\begin{aligned} S + I &\xrightarrow{\beta} I + I\\ I &\xrightarrow{\alpha} S\end{aligned}$$

is a stochastic process with non-linear transition rates in the master equation

$$\frac{d}{dt}p(I,t) = \frac{\beta}{N}(I-1)(N-(I-1))p(I-1,t) + \alpha(I+1)p(I+1,t) - \left(\frac{\beta}{N}I(N-I) + \alpha I\right)p(I,t) \quad . \tag{1.71}$$

Again, we have $S = N - I$.

For the dynamics of the mean value we obtain, with the methods as described above,

$$\frac{d}{dt}\ \langle I\rangle = (\beta-\alpha)\langle I\rangle - \frac{\beta}{N}\langle I^2\rangle \tag{1.72}$$

where now the second moment $\langle I^2\rangle := \sum_{I=1}^{N} I^2 \cdot p(I,t)$ enters the right hand side of the equation. Thus we do not obtain a closed system for the mean $\langle I\rangle$.

However, as will be described in more detail in the following chapters, an approximation, called mean field approximation, can help to close the system. It consists in

$$\langle I^2\rangle \approx \langle I\rangle^2 \tag{1.73}$$

which means essentially that the variance of the process is neglected and thus $var := \langle I^2\rangle - \langle I\rangle^2 \approx 0$. This now gives us a closed ODE,

$$\frac{d}{dt}\ \langle I\rangle = \frac{\beta}{N}\langle I\rangle(N-\langle I\rangle) - \alpha\langle I\rangle \quad . \tag{1.74}$$

Most of the considerations in this book will deal with situations where the mean field approximation only gives a first hint, but often even qualitatively fails to describe the stochastic system accurately.

Considering the density $x := \langle I\rangle/N$ instead of the absolute numbers $\langle I\rangle$ we find the simple quadratic ODE

$$\frac{dx}{dt} = \beta x(1-x) - \alpha x \quad . \tag{1.75}$$

It will appear again later in this form, e.g. as a result in Chapter 2.

1.7 Solution of the Birth–Death ODE Shows Criticality

Now we give a closer look to Eq. (1.75) and its solutions, depending on the parameter values and initial conditions. The stationary point x^* is given by the condition that the rate of change becomes zero

$$0 = \beta x^* (1 - x^*) - \alpha x^* \tag{1.76}$$

obtaining for the quadratic form in general two stationary states

$$\begin{aligned} x_1^* &= 0 \\ x_2^* &= 1 - \frac{\alpha}{\beta} \quad . \end{aligned} \tag{1.77}$$

The time solution of the ODE, Eq. (1.75), can be obtained by separation of variables and integration which gives as a result

$$x(t) = \frac{\left(1 - \frac{\alpha}{\beta}\right)}{\left(1 - e^{-(\beta-\alpha)t}\right) + \frac{1}{x_0}\left(1 - \frac{\alpha}{\beta}\right) e^{-(\beta-\alpha)t}} \tag{1.78}$$

with initial condition x_0 at starting time $t_0 = 0$.

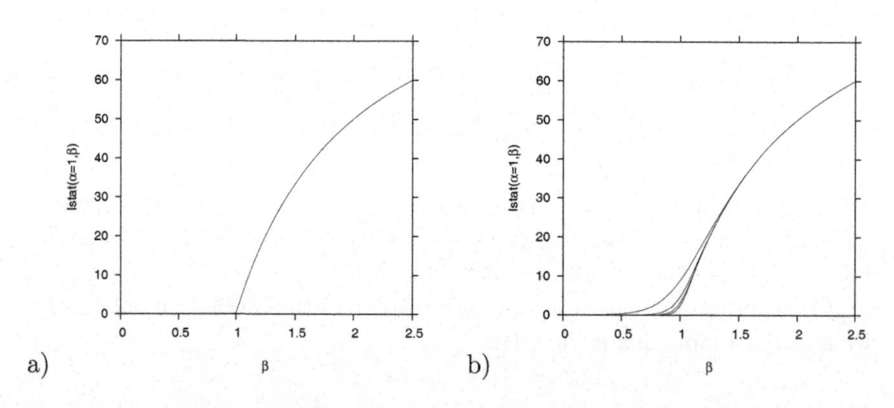

Fig. 1.4 a) Stationary states as calculated in Eq. (1.77). b) Finite time solutions from Eq. (1.78) approximate well the stationary state solutions for long enough integration times. The convergence is slowest around the critical parameter value $\beta_c = 1$.

The fixed points are approached exponentially in time

$$x(t) = \frac{\left(1 - \frac{\alpha}{\beta}\right)}{\left(1 - e^{-(\beta-\alpha)t}\right) + \frac{1}{x_0}\left(1 - \frac{\alpha}{\beta}\right) e^{-(\beta-\alpha)t}}$$

$$\longrightarrow \quad \frac{\left(1 - \frac{\alpha}{\beta}\right)}{1 - e^{-(\beta-\alpha)t}} \tag{1.79}$$

$$\sim e^{-(\beta-\alpha)t}$$

by looking at the terms which dominate the dynamic solution for large times, i.e. forgetting about the initial conditions.

However, for $\beta \to \alpha$ we have a problem with the time solution. Firstly, we see that for $\beta \to \alpha$ the second stationary point falls together with the first

$$x_2^* = 1 - \frac{\alpha}{\beta} \qquad \longrightarrow \qquad x_2^* = 0 = x_1^* \tag{1.80}$$

and for β smaller than α the solution would become negative. A stability analysis reveals that at $\beta = \alpha$ the two solutions change stability. Below, x_1^* is stable, above, x_2^* becomes stable. In that sense the point $\beta = \alpha$ marks a critical point, or β takes the critical value β_c, where in this model $\beta_c = \alpha$, which will no longer be true in spatial models.

Secondly, for $\beta \to \alpha = \beta_c$, the critical value of β, the time solution shows remarkable behavior:

$$x(t) = \frac{\left(1 - \frac{\alpha}{\beta}\right)}{1 - e^{-(\beta-\alpha)t}} \qquad \to \frac{0}{0} \quad . \tag{1.81}$$

We can, however, analyze directly in this model the solution by solving the ODE at the critical point $\beta = \beta_c$, obtaining the ODE at criticality

$$\frac{dx}{dt} = \beta x(1-x) - \alpha x = \alpha x(1-x) - \alpha x = -\alpha x^2 \quad . \tag{1.82}$$

Now this ODE $dx/dt = -\alpha x^2$ can be solved directly and gives the following result:

$$x(t) = \frac{1}{\frac{1}{x_0} + \alpha \cdot t}$$

$$\sim t^{-1} \tag{1.83}$$

which now has a power law behavior in its time dependence, as opposed to the exponential behavior in other parameter regions. The exponent -1 will turn out to be a mean field critical exponent of a whole class of stochastic systems: the directed percolation universality class. Such power law behavior will appear later in this book again, and is a general sign of behavior at and around critical states.

1.7.1 *Numerical integration shows power law at criticality*

Going back to the original variable, the mean number of infected, we have

$$I(t) = \frac{N\left(1-\frac{\alpha}{\beta}\right)}{\left(1-e^{-(\beta-\alpha)t}\right)+\frac{N}{I_0}\left(1-\frac{\alpha}{\beta}\right)e^{-(\beta-\alpha)t}} \tag{1.84}$$

and Fig. 1.5a) shows the solution for various different β values above, at and below criticality. No qualitative difference is visible between the different regimes.

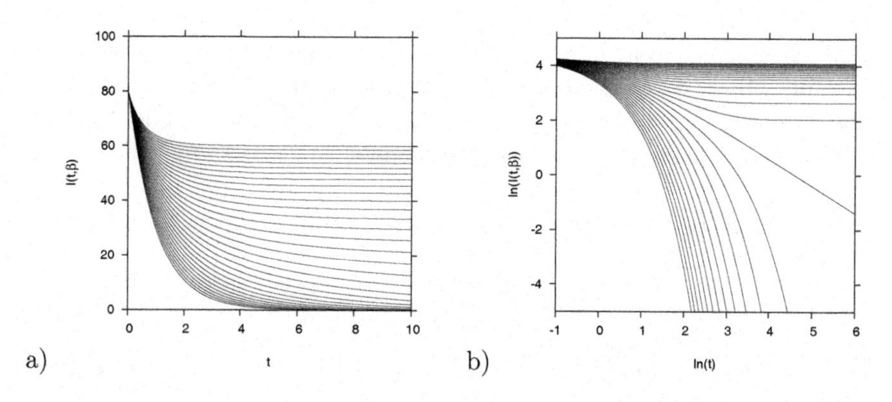

Fig. 1.5 a) For various values of β we plot the solution $I(t)$, Eq. (1.84). b) The same as in a) but in doule logarithmic plot. The critical curve shows as a straight line for large time values.

But when ploting the same as log-log plot, see Fig. 1.5b), we can clearly see the supercritical systems as moving towards finite stationary values, whereas the subcritical systems continue to decrease fast toward highly negative logarithmic values. The system at criticality $\beta = \beta_c$ is clearly visible as a straight line for larger times, separating the two regimes from each other.

When we are not plotting the total number of infected versus time, but the distance to the corresponding stationary values I^*, hence

$$I(t) - I^* = \frac{N\left(1 - \frac{\alpha}{\beta}\right)}{\left(1 - e^{-(\beta-\alpha)t}\right) + \frac{N}{I_0}\left(1 - \frac{\alpha}{\beta}\right) e^{-(\beta-\alpha)t}} - I^*_{1,2} \qquad (1.85)$$

in double logarithmic plot (Fig. 1.6) we see that above as well as below criticality, the curves converge faster to zero than the solution at criticality, visible in Fig. 1.6 as a straight line for large time values.

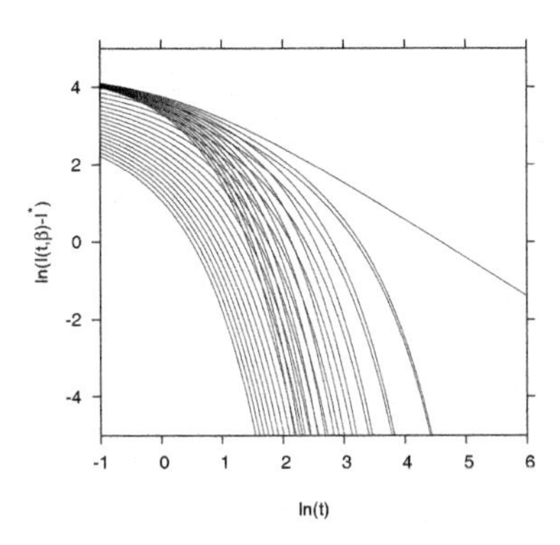

Fig. 1.6 Not $I(t)$ itself as in Fig. 1.5 but the distance of $I(t)$ to its stationary value I^*, Eq. (1.85), is plotted in double logarithmic plot. The critical curve is converging slower than the supercritical as well as the subcritical systems.

Hence at criticality the convergence to stationarity is slowed down as opposed to systems outside the critical region. This makes simulations at criticality, especially in stochastic systems as we will describe below, very time consuming. Accurate values for quantities at critcality are often very difficult to obtain.

1.7.2 *Temporal correlation length diverges at criticality*

As a final example of qualitative behavior at criticality already visible at mean field level, we show how the exponential convergence of time solutions measured by the temporal correlation length $\xi_{||}$ becomes weaker and weaker the closer we are towards the critical state. At criticality, the temporal correlation length diverges as the convergence towards stationarity is no longer exponential but slower, namely a power law again.

For long times t, the solution of Eq. (1.85) converges exponentially towards equilibrium, hence

$$I(t) - I^* = \frac{N\left(1-\frac{\alpha}{\beta}\right)}{\left(1-e^{-(\beta-\alpha)t}\right)+\frac{N}{I_0}\left(1-\frac{\alpha}{\beta}\right)e^{-(\beta-\alpha)t}} - I^*_{1,2}$$

$$\sim e^{-\frac{t}{\xi_{||}}} \quad . \tag{1.86}$$

This defines the temporal correlation length $\xi_{||}$. In Fig.1.7a) we find for large times in the semi-logarithmic plot of Eq. (1.85) straight lines for long times t, indicating the respective exponential convergence.

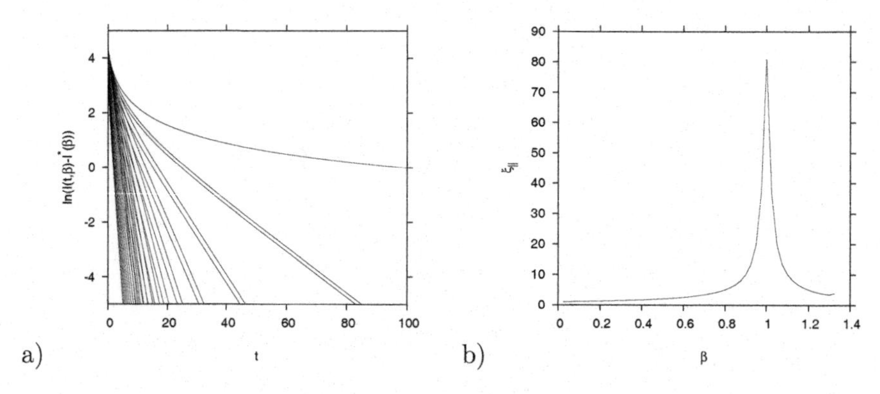

Fig. 1.7 a) The distance of $I(t)$ to its stationary value I^* for various β-values, Eq. (1.85), the same quantity as in Fig. 1.6 is now not plotted in double logarithmic plot but in semi-logarithmic plot. This shows that for large times t, the trajectories converge exponentially fast towards their equilibrium, like $e^{-t/\xi_{||}}$, hence as straight lines in this semi-logarithmic plot with slope $\xi_{||}$. b) The plot of the final slope $\xi_{||}$, taken from a) versus the parameter value β is shown. The slow convergence to equilibrium at criticality, $\beta = \beta_c = 1$, can be observed as a divergent peak of $\xi_{||}(\beta)$ around $\beta = 1$.

The convergence is slowest near criticality, and in this region the scaling is given by the expression

$$\xi_{||} \sim |\beta - \beta_c|^{-\nu_{||}} \quad . \tag{1.87}$$

Hence with a power law behavior, and in mean field approximation, we have the exponent $\nu_{||} = 1$. Fig. 1.7b) demonstrates the divergence of the temporal correlation length at criticality. For non-mean field models we will calculate the respective quantities in Chapter 5.

Chapter 2

Spatial Stochastic Birth–Death Process or SIS-Epidemics

In Chapter 1 we mentioned that the birth and death process results in a non-linear master equation. In order to better understand the quite complicated transition rates we now introduce the spatial stochastic birth–death process, where the transition rates come rather more naturally. However, there is some cost to it; namely that the probability distribution acts on many variables, as many as individuals or spaces are described. Again, it will emerge that moment equations can be derived as in Chapter 1, where the term "mean field approximation" will also become easier to understand, as the local variance will be neglected.

2.1 The Spatial Master Equation

One of the simplest and best studied spatial processes is the birth–death process, with birth rate b and death rate a on N sites. A site i can either have an individual $I_i := 1$ or be an empty space $S_i := 1$, hence $I_i = 0$ (in general $S_i := 1 - I_i$). We use I for an individual, and S for a space being empty. Translated into epidemiology, I is the infected, S the susceptible class, b the infection rate and a the recovery. We refer to it as the SIS-system.

The master equation for the spatial SIS-system for N lattice points is

$$\begin{aligned} \frac{d}{dt}p(I_1, ..., I_N, t) = \sum_{i=1}^{N} & w_{I_i,1-I_i} \; p(I_1, ..., 1 - I_i, ..., I_N, t) \\ & - \sum_{i=1}^{N} w_{1-I_i,I_i} \; p(I_1, ..., I_i, ..., I_N, t) \end{aligned} \tag{2.1}$$

for $I_i \in \{0,1\}$ and transition rate

$$w_{I_i,1-I_i} = b \left(\sum_{j=1}^{N} J_{ij} I_j \right) \cdot I_i + a \cdot (1 - I_i) \quad , \tag{2.2}$$

and

$$w_{1-I_i,I_i} = b \left(\sum_{j=1}^{N} J_{ij} I_j \right) \cdot (1 - I_i) + a \cdot I_i \quad , \tag{2.3}$$

with birth or infection rate b and death or recovery rate a. Here (J_{ij}) is the adjacency matrix containing 0 for no connection and 1 for a connection between sites i and j. Hence $J_{ij} = J_{ji} \in \{0,1\}$ for $i \neq j$ and $J_{ii} = 0$. Thus a state is now defined by $I_1, ..., I_N$, and the probability of each state has to be considered to describe the spatial system accurately.

For the epidemiological example, the SIS epidemics, we have the following scenario: looking at one site i, the transition rate b gives the probability per time of infection by a neighboring site j. The number of infected sites that are neighbors to site i is given by $\sum_{j=1}^{N} J_{ij} I_j$, so that this time b gives the force of infection, $b \cdot \left(\sum_{j=1}^{N} J_{ij} I_j \right)$, to site i. Since you can only become infected, $I_i = 1$, if you were susceptible before, $S_i = 1 = 1 - I_i$, the transition rate $w_{I_i,1-I_i}$ only leads to a new infection if $1 - I_i = 0$, hence $I_i = 1$. This is why we have to multiply the force of infection $b \cdot \left(\sum_{j=1}^{N} J_{ij} I_j \right)$ with I_i in the transition rate entering the state with I_i, i.e. $w_{I_i,1-I_i}$, and multiply by $1 - I_i$ in the transition rate leaving the state with I_i, i.e. w_{1-I_i,I_i}. A simpler reasoning along the same lines holds also for the recovery terms, including a.

For the birth–death example, individuals I_j have a probability per time to give birth to empty spaces S_i with rate b. The rest of the above holds again. Additionally, an individual I_i can die with rate a.

We use the master equation approach for a spatial system as for example used in [Glauber (1963)] for a spin system, as we will see in Chapter 3 in more detail.

2.1.1 *A first inspection of the spatial birth–death process*

Figure 2.1 shows simulations of the one dimensional birth and death process for two parameter sets. The horizontal axis gives the one dimensional space, while time goes from top to bottom. For low death rates or high birth rates we see that the system approaches quickly the stationary state and then just shows some fast changing noise around that state (see Fig. 2.1a)).

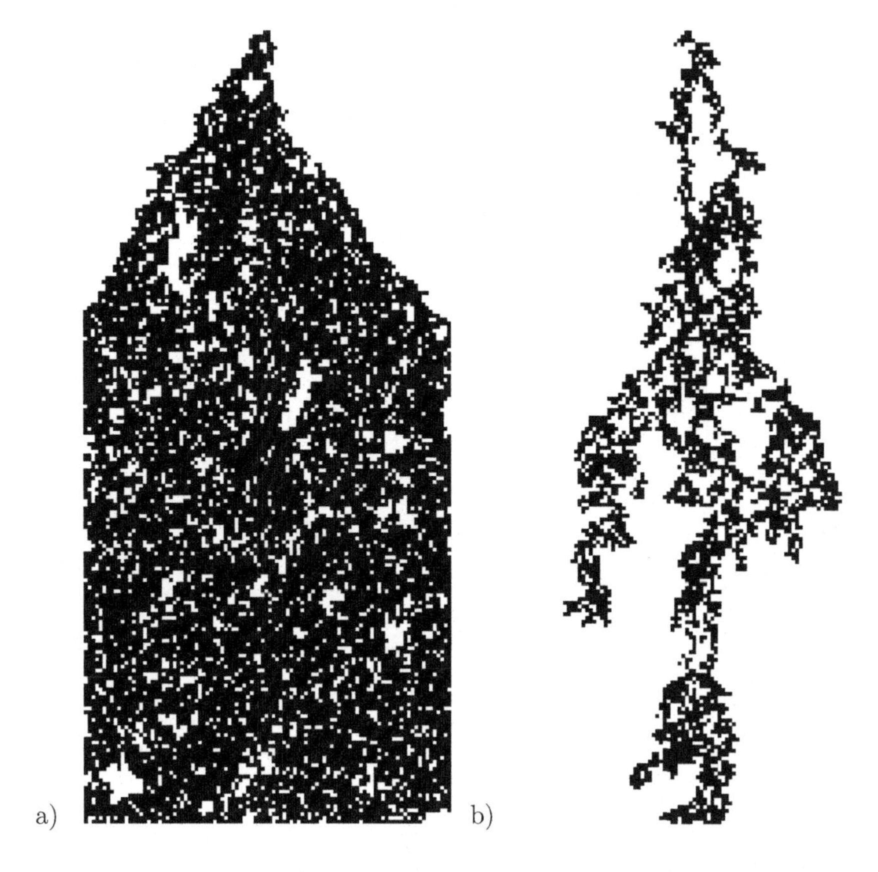

Fig. 2.1 One dimensional birth–death process with $N = 100$ individuals. Parameters: $b = 1$ fixed, and a varied, a) $a = 0.3$, low death rate gives huge population level, b) $a = 0.62$. Space goes horizontally, time from top to bottom.

However, for an increasing death rate, respectively a decreasing birth rate, the stationary state is lower, but also is approached more slowly (see Fig. 2.1b)). For a low stationary state in particular we observe huge fluctuations around that stationary state, also with much longer autocorrelation.

For even higher death rates, we observe a further increase in fluctuations with even longer autocorrelation, eventually leading to the extinction of the process. Of course, for very high death rates, respectively low birth rates, the process tends to die out very quickly after some initial fluctuations.

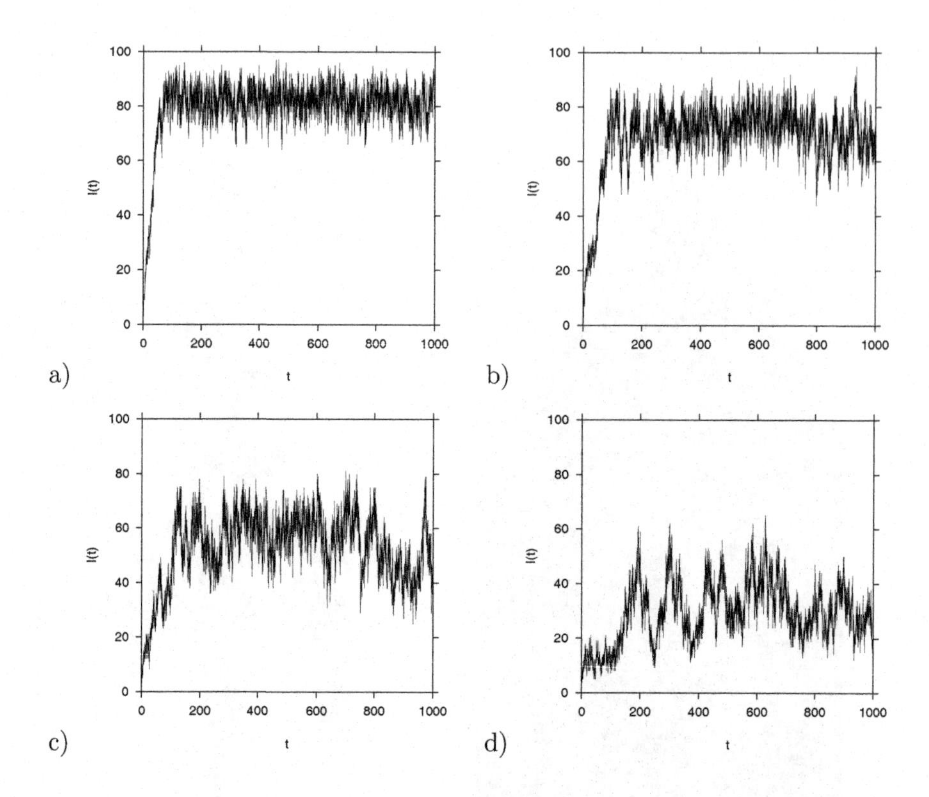

Fig. 2.2 One dimensional birth–death process with $N = 100$ individuals. Parameters: $b = 1$ fixed, and a varied, a) $a = 0.3$, low death rate gives huge population level, b) $a = 0.4$, c) $a = 0.5$, d) $a = 0.6$. High death rate gives not only a smaller mean population level but also larger variance.

This becomes clearer in the time series plots of the system in Figs. 2.2 and 2.3. For the different panels we decrease the death rate, observing large fluctuations but remaining far away from the extinction boundary in Fig. 2.2d) for $a = 0.6$. In Fig. 2.3a), for $a = 0.62$ we find the system at times close to extinction and due to stochastic events it might easily become extinct. In Fig. 2.3b) for $a = 0.63$ we actually observe in this given realization of the stochastic process an extinction around time 900, whereas for lower values of a extinction rapidly happens.

A closer examination of the numerical values of the simulations shows us that the region of huge fluctuations to quick extinction is rather narrow in parameter space. This is an indication of the existance of a critical threshold.

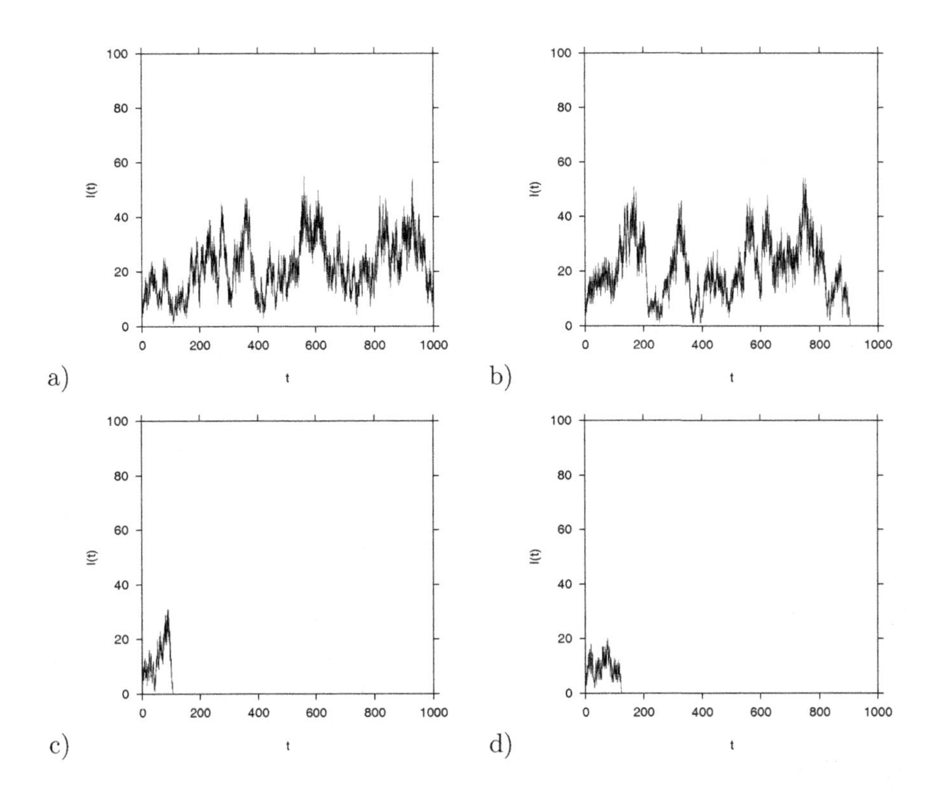

Fig. 2.3 Around criticality, a) $a = 0.62$, b) $a = 0.63$, c) $a = 0.64$, d) $a = 0.7$. For an even higher death rate the process dies out quickly, and variance is reduced again.

2.2 Clusters and their Dynamics

Now, we want to describe the stochastic system using easily accessible global quantities, such as the dynamics of the total number of individuals, or the number of clusters of certain shapes. Since the dynamics of the total number of individuals already will contain neighboring pairs, due to the non-linearity in the transition rates e.g. $w_{1-I_i,I_i} \sim I_i \cdot I_j$, we need to look at clusters anyway. The methods we use here are analogously with the methods used for the non-spatial master equations in Chapter 1.

Defining the number of clusters with certain shapes, for the total number of individuals we have

$$[I] := \sum_{i=1}^{N} I_i \tag{2.4}$$

and respectively

$$[S] := \sum_{i=1}^{N} (1 - I_i) \quad . \tag{2.5}$$

For pairs we have

$$[II] := \sum_{i=1}^{N} \sum_{j=1}^{N} J_{ij} \, I_i \cdot I_j \tag{2.6}$$

and triples

$$[III] := \sum_{i=1}^{N} \sum_{j=1}^{N} \sum_{k=1}^{N} J_{ij} J_{jk} \cdot I_i I_j I_k \tag{2.7}$$

or triangles

$$[\Delta] := \sum_{i=1}^{N} \sum_{j=1}^{N} \sum_{k=1}^{N} J_{ij} J_{jk} J_{ki} \cdot I_i I_j I_k \tag{2.8}$$

and so on. These space averages, e.g. $[I] := \sum_{i=1}^{N} I_i$, depend on the ensemble $(I_1, ..., I_N)$ which changes with time. Hence we define the ensemble average, e.g.

$$\langle I \rangle(t) := \sum_{I_1=0}^{1} ... \sum_{I_N=0}^{1} [I] \, p(I_1, ..., I_N, t)$$

or more generally for any function $f = f(I_1, ..., I_N)$ of the state variables we define the ensemble average as

$$\langle f \rangle(t) := \sum_{I_1=0}^{1} ... \sum_{I_N=0}^{1} f(I_1, ..., I_N) \, p(I_1, ..., I_N, t) \quad . \tag{2.9}$$

We will consider mainly functions like $f = [I]$, $f = [II]$ etc. Then, the time evolution is determined by

$$\frac{d}{dt} \langle f \rangle(t) := \sum_{I_1=0}^{1} ... \sum_{I_N=0}^{1} f(I_1, ..., I_N) \, \frac{d}{dt} p(I_1, ..., I_N, t) \tag{2.10}$$

where the master equation is to be inserted again, as shown in Chapter 1, giving terms of the form $\langle f \rangle$ and other expressions $\langle g(I_1, ..., I_N) \rangle$. By defining marginal distributions

$$p(I_i, t) := \sum_{I_1=0}^{1} \dots \not\sum_{I_i=0}^{1} \dots \sum_{I_N=0}^{1} p(I_1, ..., I_N, t) \tag{2.11}$$

and respectively

$$p(I_i, I_j, t) := \sum_{I_1=0}^{1} \dots \not\sum_{I_i=0}^{1} \dots \not\sum_{I_j=0}^{1} \dots \sum_{I_N=0}^{1} p(I_1, ..., I_N, t) \tag{2.12}$$

one obtains for its realizations useful expressions like

$$\begin{aligned} p(I_i\!=\!1, I_j\!=\!0, t) &:= \sum_{I_1=0}^{1} \dots \not\sum_{I_i=0}^{1} \dots \not\sum_{I_j=0}^{1} \dots \sum_{I_N=0}^{1} p(I_1, ..., I_i\!=\!1, I_j\!=\!0, ..., I_N, t) \\ &= \sum_{I_1=0}^{1} \dots \sum_{I_N=0}^{1} I_i(1 - I_j)\, p(I_1, ..., I_N, t) \\ &=: \langle I_i(1 - I_j) \rangle = \langle I_i \rangle - \langle I_i I_j \rangle \end{aligned}$$

which we will consider extensively in the subsequent text. Hence, it follows that for the total number of pairs we have

$$\langle II \rangle(t) = \sum_{i=1}^{N} \sum_{j=1}^{N} J_{ij} \langle I_i I_j \rangle \tag{2.13}$$

and with $\langle S_i I_j \rangle = \langle (1 - I_i) I_j \rangle$ also

$$\langle SI \rangle(t) = \sum_{i=1}^{N} \sum_{j=1}^{N} J_{ij} \langle S_i I_j \rangle = \sum_{i=1}^{N} \langle I_i \rangle \left(\sum_{j=1}^{N} J_{ij} \right) - \sum_{i=1}^{N} \sum_{j=1}^{N} J_{ij} \langle I_i I_j \rangle \tag{2.14}$$

with

$$\sum_{i=1}^{N} \langle I_i \rangle \left(\sum_{j=1}^{N} J_{ij} \right) = Q \cdot \sum_{i=1}^{N} \langle I_i \rangle = Q \cdot \langle I \rangle \tag{2.15}$$

for $Q_i := \left(\sum_{j=1}^{N} J_{ij} \right)$ the number of neighbors to site i. Here we assume the Q_i to be constant $Q_i = Q$ for all lattice sites i, since we are mainly interested in regular lattices (and have to assume even periodic boundary conditions). For irregular or random lattices the index i has to be kept for

Q_i. Here, a lot more complexity is possible in irregular lattices as opposed to the regular ones considered in the following.

In general, terms of the form

$$\langle II \rangle_\nu := \sum_{i=1}^{N} \sum_{j=1}^{N} J_{ij}^\nu \cdot I_i I_j \tag{2.16}$$

will appear with any ν^{th} power of the adjacency matrix, e.g. $J_{ij}^2 = \sum_{k=1}^{N} J_{ik} J_{kj}$, and respectively

$$\langle III \rangle_{\mu,\nu} := \sum_{i=1}^{N} \sum_{j=1}^{N} \sum_{k=1}^{N} J_{ij}^\mu J_{jk}^\nu \cdot I_i I_j I_k \tag{2.17}$$

and so on.

2.2.1 *Time evolution of marginals and local expectations*

For the marginals we can put forward some rules which are rigorous but also intuitively obtained from the master equation.

The birth–death process (or equivalently the SIS-epidemics, and for a more general class of processes specified below) presents the following expressions for the dynamics of local quantities such as $\langle I_i \rangle$ in the following way

$$\begin{aligned} \frac{d}{dt} \langle I_i \rangle &:= \sum_{I_1=0}^{1} \dots \sum_{I_N=0}^{1} I_i \ \frac{d}{dt} p(I_1, ..., I_N, t) \\ &= \sum_{\{I\}} I_i \left(\sum_{k=1}^{N} w_{I_k, 1-I_k} \ p(I_1, ..., 1-I_k, ..., I_N, t) \right. \\ &\quad \left. - \sum_{k=1}^{N} w_{1-I_k, I_k} \ p(I_1, ..., I_k, ..., I_N, t) \right) \quad . \end{aligned} \tag{2.18}$$

This uses the definition $\sum_{\{I\}} := \sum_{I_1=0}^{1} \cdots \sum_{I_N=0}^{1}$ for the ensemble average and inserts the master equation for the time derivative of the probability.

The following calculation rule for any function $f(I_i, I_j)$

$$\sum_{\{I\}} f(I_i, I_j)\, p(I_1, ..., 1-I_i, ..., I_N, t) = \sum_{\{I\}} f(1-I_i, I_j)\, p(I_1, ..., I_i, ..., I_N, t) \tag{2.19}$$

from the elementary consideration $\sum_{I_i=0}^{1} f(I_i)p(1-I_i,t) = f(0)p(1,t) + f(1)p(0,t) = f(1)p(0,t) + f(0)p(1,t) = \sum_{I_i=0}^{1} f(1-I_i)p(I_i,t)$ results in

$$\begin{aligned}\frac{d}{dt}\langle I_i\rangle = \sum_{\{I\}} I_i \Bigg(& \sum_{k=1,k\neq i}^{N} w_{1-I_k,I_k}\ p(I_1,...,I_k,...,I_N,t) \\ & - \sum_{k=1,k\neq i}^{N} w_{1-I_k,I_k}\ p(I_1,...,I_k,...,I_N,t)\Bigg) \\ & + \sum_{\{I\}} (1-I_i)\ w_{1-I_i,I_i}\ p(I_1,...,I_i,...,I_N,t) \\ & -I_i\ w_{1-I_i,I_i}\ p(I_1,...,I_i,...,I_N,t)) \\ = & \sum_{\{I\}} w_{1-I_i,I_i}\ \Big((1-I_i)-I_i\Big)\ p(I_1,...,I_i,...,I_N,t) \quad .\end{aligned} \tag{2.20}$$

For the variable $I_i \in \{0,1\}$ we obtain the equations $I_i^2 = I_i$ and $(1-I_i)^2 = (1-I_i)$ and hence $I_i(1-I_i) = 0$, so that for the birth–death process

$$\begin{aligned} w_{1-I_i,I_i}\cdot\Big((1-I_i)-I_i\Big) &= b\left(\sum_{j=1}^{N} J_{ij}I_j\right)\cdot\underbrace{\Big((1-I_i)^2-(1-I_i)I_i\Big)}_{=(1-I_i)} \\ & \quad +a\underbrace{\Big(I_i(1-I_i)-I_i^2\Big)}_{=-I_i} \\ &=: \tilde{w}_{1-I_i,I_i} \end{aligned} \tag{2.21}$$

with a function $\tilde{w}$ with additive birth and subtractive death term. The equation

$$w_{1-I_i,I_i}\cdot\Big((1-I_i)-I_i\Big) = \tilde{w}_{1-I_i,I_i} \tag{2.22}$$

holds for general transition probabilities of the functional form

$$w_{1-I_i,I_i} = f(\{I_j\}_{j\neq i})\cdot(1-I_i) + g(\{I_j\}_{j\neq i})\cdot I_i \tag{2.23}$$

with arbitrary functions f for birth terms and g for death terms and $\tilde{w}$ defined as

$$\tilde{w}_{1-I_i,I_i} := f(\{I_j\}_{j\neq i}) \cdot (1-I_i) - g(\{I_j\}_{j\neq i}) \cdot I_i \quad . \tag{2.24}$$

Hence, we obtain

$$\begin{aligned}\frac{d}{dt}\langle I_i\rangle &= \sum_{\{I\}} \tilde{w}_{1-I_i,I_i}\; p(I_1,...,I_i,...,I_N,t) \\ &= b\sum_{j=1}^{N} J_{ij}\langle I_j(1-I_i)\rangle - a\langle I_i\rangle \\ &= b\sum_{j=1}^{N} J_{ij}\langle S_i I_j\rangle - a\langle I_i\rangle\end{aligned} \tag{2.25}$$

which provides an easy and intuitive way to calculate generally dynamics of local expectation values. In the last line we used again $S_i := 1 - I_i$.

2.3 Moment Equations

For the total number $\langle I\rangle := \sum_{i=1}^{N}\langle I_i\rangle$ we obtain the dynamics

$$\begin{aligned}\frac{d}{dt}\langle I\rangle &= \sum_{i=1}^{N}\frac{d}{dt}\langle I_i\rangle \\ &= \sum_{i=1}^{N}\left(-a\langle I_i\rangle + b\sum_{j=1}^{N} J_{ij}(\langle I_j\rangle - \langle I_i I_j\rangle)\right)\end{aligned} \tag{2.26}$$

and in detail

$$\begin{aligned}\frac{d}{dt}\langle I\rangle &= -a\underbrace{\sum_{i=1}^{N}\langle I_i\rangle}_{=\langle I\rangle} + \underbrace{b\sum_{j=1}^{N}\langle I_j\rangle\underbrace{\sum_{i=1}^{N} J_{ij}}_{=Q_j=Q}}_{=Q\langle I\rangle} - b\underbrace{\sum_{i=1}^{N}\sum_{j=1}^{N} J_{ij}\langle I_i I_j\rangle}_{=\langle II\rangle_1} \\ &= -a\langle I\rangle + bQ\langle I\rangle - b\langle II\rangle_1\end{aligned}$$

such that

$$\frac{d}{dt}\langle I\rangle = b\Big(Q\langle I\rangle - \langle II\rangle_1\Big) - a\langle I\rangle$$
$$= b\langle SI\rangle_1 - a\langle I\rangle \tag{2.27}$$

with $\langle SI\rangle_1 := \sum_{i=1}^{N}\sum_{j=1}^{N} J_{ij}\langle S_i I_j\rangle = Q\langle I\rangle - \langle II\rangle_1$. To obtain the dynamics for the total number of pairs

$$\frac{d}{dt}\langle II\rangle_1 = \sum_{i=1}^{N}\sum_{j=1}^{N} J_{ij}\frac{d}{dt}\langle I_i I_j\rangle \tag{2.28}$$

we first have to calculate $\frac{d}{dt}\langle I_i I_j\rangle$ from the rules given above and insert the spatial master equation. It is explicitly

$$\begin{aligned}\frac{d\langle I_i I_j\rangle}{dt} &= \sum_{\{I\}} I_i I_j \,\frac{d}{dt}\, p(I_1,...,I_N,t)\\ &= \sum_{\{I\}} \underbrace{\Big(I_j((1-I_i)-I_i)w_{1-I_i,I_i} + I_i((1-I_j)-I_j)w_{1-I_j,I_j}\Big)}_{\text{see Eq. (2.20)}} p(I_1,...,I_N)\\ &= \underbrace{\langle I_j\tilde{w}_{1-I_i,I_i} + I_i\tilde{w}_{1-I_j,I_j}\rangle}_{\text{see Eq. (2.21)}}\end{aligned}$$

and

$$\begin{aligned}\frac{d\langle I_i I_j\rangle}{dt} &= \langle I_j(b\sum_{k=1}^{N} J_{ik}I_k(1-I_i) - aI_i)\rangle + \langle I_i(b\sum_{k=1}^{N} J_{jk}I_k(1-I_j) - aI_j)\rangle\\ &= b\sum_{k=1}^{N} J_{ik}\langle I_j I_k(1-I_i)\rangle - a\langle I_j I_i\rangle + b\sum_{k=1}^{N} J_{jk}\langle I_i I_k(1-I_j)\rangle - a\langle I_i I_j\rangle\ .\end{aligned}$$

Hence for the dynamics of nearest neighbor pairs we obtain

$$\begin{aligned}\frac{d}{dt}\langle II\rangle_1 &= \sum_{i=1}^{N}\sum_{j=1}^{N} J_{ij}\frac{d}{dt}\langle I_i I_j\rangle \\ &= -2a\sum_{i=1}^{N}\sum_{j=1}^{N} J_{ij}\langle I_i I_j\rangle - 2b\sum_{i=1}^{N}\sum_{j=1}^{N}\sum_{k=1}^{N} J_{ij}\, J_{jk}\langle I_i I_j I_k\rangle \\ &\quad +2b\sum_{i=1}^{N}\sum_{k=1}^{N}\underbrace{\left(\sum_{j=1}^{N} J_{ij}\, J_{jk}\right)}_{=:(J^2)_{ik}}\langle I_i I_k\rangle \quad .\end{aligned}$$

Here $(J^2)_{ij}$ is the matrix J squared and then taken the ij th element of that matrix J^2. This last term gives a contribution of the form $\langle II\rangle_2$ (see Eq. (2.16)).

In total, we obtain for the pair dynamics

$$\begin{aligned}\frac{d}{dt}\langle II\rangle_1 &= 2b\left(\langle II\rangle_2 - \langle III\rangle_{1,1}\right) - 2a\langle II\rangle_1 \\ &= 2b\langle ISI\rangle_{1,1} - 2a\langle II\rangle_1\end{aligned} \tag{2.29}$$

with $\langle ISI\rangle_{1,1} := \sum_{i=1}^{N}\sum_{j=1}^{N}\sum_{k=1}^{N} J_{ij}J_{jk}\langle I_i(1-I_j)I_k\rangle$. Again, the ODE for the nearest neighbor's pair $\langle II\rangle_1$ involves higher moment terms like $\langle II\rangle_2$ and $\langle III\rangle_{1,1}$, as we have seen in the non-spatial case in Chapter 1.

We now try to approximate the higher moments in terms of lower ones in order to close the ODE system. The quality of the approximation will depend on the actual parameters of the birth-death process, i.e. a and b. We first investigate the mean field approximation, expressing $\langle II\rangle_1$ in terms of $\langle I\rangle$. Then other schemes to approximate higher moments are shown, such as the pair approximation.

2.3.1 *Mean field behavior*

In mean field approximation, in the interaction term the exact number of inhabited neighbors is replaced by the average number of inhabitants in the full system, thus acting like a mean field on the actually considered site.

Hence we set

$$\sum_{j=1}^{N} J_{kj} I_j \approx \sum_{j=1}^{N} J_{kj} \frac{\langle I \rangle}{N}$$
$$= \frac{Q}{N} \cdot \langle I \rangle \quad , \tag{2.30}$$

where the last line of Eq. (2.30) only holds again for regular lattices. We get for $\langle II \rangle_1$ in Eq. (2.26)

$$\langle II \rangle_1 = \langle \sum_{i=1}^{N} \sum_{j=1}^{N} J_{ij} I_i I_j \rangle = \langle \sum_{i=1}^{N} I_i \sum_{j=1}^{N} J_{ij} I_j \rangle$$
$$\approx \langle \sum_{i=1}^{N} I_i \frac{Q}{N} \cdot \langle I \rangle \rangle = \frac{Q}{N} \cdot \langle I \rangle \cdot \langle \sum_{i=1}^{N} I_i \rangle \tag{2.31}$$
$$= \frac{Q}{N} \cdot \langle I \rangle^2 \quad .$$

More formally we can investigate the local variance, see e.g. [Greiner, Neise and Stöcker (1987), pp. 463–470], to obtain the mean field approximation. This gives here the same result as the one just shown, but is sometimes more accurate, see Section 3.4. The local variance is

$$\left(I_i - \frac{\langle I \rangle}{N} \right) \left(I_j - \frac{\langle I \rangle}{N} \right) = I_i I_j - I_i \frac{\langle I \rangle}{N} - I_j \frac{\langle I \rangle}{N} + \frac{1}{N^2} \langle I \rangle^2 \quad . \tag{2.32}$$

Hence the pair gives

$$I_i I_j = I_i \frac{\langle I \rangle}{N} + I_j \frac{\langle I \rangle}{N} - \frac{1}{N^2} \langle I \rangle^2 + \underbrace{\left(I_i - \frac{\langle I \rangle}{N} \right) \left(I_j - \frac{\langle I \rangle}{N} \right)}_{\approx 0} \tag{2.33}$$

which means that we neglect the local variance in mean field approximation. Thus, for the interaction term we find that

$$\begin{aligned}\sum_{i=1}^{N}\sum_{j=1}^{N} J_{ij}I_iI_j &\approx \sum_{i=1}^{N}\sum_{j=1}^{N} J_{ij}\left(I_i\frac{\langle I\rangle}{N} + I_j\frac{\langle I\rangle}{N} - \frac{1}{N^2}\langle I\rangle^2\right) \\ &= \frac{\langle I\rangle}{N}\left(\sum_{i=1}^{N} I_i \underbrace{\sum_{j=1}^{N} J_{ij}}_{=Q} + \sum_{j=1}^{N} I_j \sum_{i=1}^{N} J_{ij}\right) \\ &\quad -\frac{1}{N^2}\langle I\rangle^2 \underbrace{\left(\sum_{i=1}^{N}\sum_{j=1}^{N} J_{ij}\right)}_{=Q\sum_{i=1}^{N} 1 = NQ}\end{aligned} \tag{2.34}$$

and cleaning up gives

$$\begin{aligned}\sum_{i=1}^{N}\sum_{j=1}^{N} J_{ij}I_iI_j &= 2\frac{Q}{N}\langle I\rangle\left(\sum_{i=1}^{N} I_i\right) - \frac{1}{N^2}\langle I\rangle^2 NQ \\ &= \frac{Q}{N}\left(2\langle I\rangle\cdot\left(\sum_{i=1}^{N} I_i\right) - \langle I\rangle^2\right)\end{aligned} \tag{2.35}$$

such that

$$\langle\sum_{i=1}^{N}\sum_{j=1}^{N} J_{ij}I_iI_j\rangle = \frac{Q}{N}\left(2\langle I\rangle\cdot\langle I\rangle - \langle I\rangle^2\right) = \frac{Q}{N}\langle I\rangle^2 \tag{2.36}$$

hence the same result as in the simpler way, which was shown before, to describe the mean field assumption. Thus we obtain the result of the dynamics for the total mean number of individuals in mean field approximation:

$$\begin{aligned}\frac{d}{dt}\langle I\rangle &= b\left(Q\langle I\rangle - \frac{Q}{N}\langle I\rangle^2\right) - a\langle I\rangle \\ &= b\frac{Q}{N}\left(N - \langle I\rangle\right)\langle I\rangle - a\langle I\rangle \quad .\end{aligned} \tag{2.37}$$

For homogeneous mixing, i.e. where the number of neighbors equals the total population size $Q \approx N$, we obtain the logistic equation for the total number of inhabited sites

$$\frac{d}{dt} \langle I \rangle = b \ \langle I \rangle (N - \langle I \rangle) - a \langle I \rangle \tag{2.38}$$

or for the proportion $\frac{\langle I \rangle}{N} =: x \in [0, 1]$

$$\frac{d}{dt} \frac{\langle I \rangle}{N} = Nb \ \frac{\langle I \rangle}{N} \left(1 - \frac{\langle I \rangle}{N} \right) - a \frac{\langle I \rangle}{N} \quad . \tag{2.39}$$

Hence

$$\frac{dx}{dt} = Nb \ x \cdot (1 - x) - a \cdot x \quad . \tag{2.40}$$

This is again the logistic equation as seen in Chapter 1. We will now go beyond the mean field approximation and approximate triples into pairs in the next two sections, showing the analytic steps and and assumptions that are required.

2.3.2 *Pair approximation*

From the general scheme, say for 3 neighbors, S, I, and R triples can be approximated by pairs and singles, see e.g. [Rand (1999)], in the following way:

$$\langle SIR \rangle \approx \frac{\langle SI \rangle \cdot \langle IR \rangle}{\langle I \rangle} \quad . \tag{2.41}$$

To approximate triples into pairs and simpler, we first formulate and prove an analogue to Bayes' formula for local pairs $\langle I_i I_j \rangle = \langle I_i | I_j = 1 \rangle \cdot \langle I_j \rangle$, which then leads, under some approximation, to a local analogue to the pair approximation $\langle I_i I_j I_k \rangle \approx \frac{\langle I_i I_j \rangle}{\langle I_j \rangle} \cdot \langle I_j I_k \rangle$. We then generalize to the total number of triples to be approximated by pairs. Bayes' analogue for the local mean is

$$\langle I_i I_j \rangle := \sum_{I_i=0}^{1} \sum_{I_j=0}^{1} I_i I_j \underbrace{p(I_i, I_j)}_{\substack{=p(I_i|I_j)\cdot p(I_j) \\ \text{Bayes' rule}}} = \sum_{I_j=0}^{1} \underbrace{\left(\sum_{I_i=0}^{1} I_i \, p(I_i|I_j) \right)}_{\substack{=:\langle I_i|I_j \rangle \\ \text{conditional mean}}} I_j p(I_j) \tag{2.42}$$

defining the conditional mean. With this it is

$$\langle I_i I_j \rangle = \underbrace{\langle I_i|I_j=0\rangle \cdot 0 \cdot p(I_j=0)}_{=0} + \langle I_i|I_j=1\rangle \cdot 1 \cdot p(I_j=1)$$

$$= \langle I_i|I_j=1\rangle \cdot \underbrace{\left(1 \cdot p(I_j=1) + \underbrace{0 \cdot p(I_j=1)}_{=0} \right)}_{=\sum_{I_j=0}^{1} I_j p(I_j) = \langle I_j \rangle}$$

such that in total we obtain

$$\langle I_i I_j \rangle = \langle I_i|I_j = 1\rangle \cdot \langle I_j \rangle \quad . \tag{2.43}$$

From the derivation we also see that this formula only holds for the zero-one valued variable we consider here, $I_i \in \{0, 1\}$.

Respectively, for local triples we obtain, using the same arguments,

$$\langle I_i I_j I_k \rangle = \langle I_i|I_j=1, I_k=1\rangle \cdot \langle I_j I_k \rangle \tag{2.44}$$

since for the zero-one valued variable it is

$$\sum_{I_j=0}^{1} \sum_{I_k=0}^{1} I_j I_k \, p(I_j, I_k) = 1 \cdot 1 \cdot p(I_j=1, I_k=1) \quad . \tag{2.45}$$

Now, if we only have weak stochastic dependencies, hence the variable I_i only depends on its nearest neighbor I_j, but not on neighbors of I_j, we can approximate

$$\langle I_i | I_j = 1, I_k = 1 \rangle \approx \langle I_i | I_j = 1 \rangle \tag{2.46}$$

and then insert Bayes' analogue for pairs in the form $\langle I_i | I_j = 1 \rangle = \frac{\langle I_i I_j \rangle}{\langle I_i \rangle}$ such that we finally obtain the pair approximation of triples for local variables

$$\langle I_i I_j I_k \rangle \approx \frac{\langle I_i I_j \rangle}{\langle I_i \rangle} \cdot \langle I_j I_k \rangle \quad . \tag{2.47}$$

This, of course, will only give good numerical results when we have site j as neighbor to i, and k as neighbor to j, but not to i again, hence when triangles are absent or rare in our networks.

To generalize Eq. (2.47) to the global quantity of the total number of triples $\langle III \rangle_{1,1} := \sum_{i=1}^{N} \sum_{j=1}^{N} \sum_{k=1}^{N} J_{ij} J_{jk} \langle I_i I_j I_k \rangle$, we have to make some further approximating considerations, namely for the quantity in the calculations appearing quantity

$$\sum_{i=1}^{N} J_{ij} \langle I_i | I_j = 1 \rangle =: \langle I | I_{\text{nn},j} = 1 \rangle \approx \langle I | I_{\text{nn}} = 1 \rangle \tag{2.48}$$

which appears in the calculations frequently. Here, $\langle I | I_{\text{nn},j} = 1 \rangle$ is the mean of the total number of I to which j is neighbor with value $I_j = 1$, and then the approximation that this quantity becomes independent of site j, $\langle I | I_{\text{nn},j} = 1 \rangle \approx \langle I | I_{\text{nn}} = 1 \rangle$.

Hence for the total number of triples we have the approximation

$$\begin{aligned} \langle III \rangle_{1,1} &:= \sum_{i=1}^{N} \sum_{j=1}^{N} \sum_{k=1}^{N} J_{ij} J_{jk} \underbrace{\langle I_i I_j I_k \rangle} \\ &\qquad = \underbrace{\langle I_i | I_j = 1, I_k = 1 \rangle}_{\approx \langle I_i | I_j = 1 \rangle} \cdot \langle I_i | I_j \rangle \\ &\qquad \text{independence of next nearest neighbors} \\ &\approx \sum_{j=1}^{N} \sum_{k=1}^{N} J_{jk} \underbrace{\left(\sum_{i=1}^{N} J_{ij} \langle I_i | I_j = 1 \rangle \right)}_{=: \langle I | I_{\text{nn},j} = 1 \rangle} \cdot \langle I_j I_k \rangle \\ &\qquad \text{mean conditioned on nearest neighbors} \end{aligned} \tag{2.49}$$

which gives again with another approximation

$$\langle III\rangle_{1,1} = \sum_{j=1}^{N}\sum_{k=1}^{N} J_{jk} \underbrace{\langle I|I_{\mathrm{nn},j}=1\rangle}_{\approx\langle I|I_{\mathrm{nn}}=1\rangle} \cdot \langle I_j I_k\rangle$$

$$\approx \langle I|I_{\mathrm{nn}}=1\rangle \cdot \underbrace{\sum_{j=1}^{N}\sum_{k=1}^{N} J_{jk}\langle I_j I_k\rangle}_{\langle II\rangle_1}$$

$$= \langle I|I_{\mathrm{nn}}=1\rangle \cdot \langle II\rangle_1$$

such that in total we obtain

$$\langle III\rangle_{1,1} \approx \langle I|I_{\mathrm{nn}}=1\rangle \cdot \langle II\rangle_1 \tag{2.50}$$

with $\langle I|I_{\mathrm{nn}}=1\rangle$ still to be evaluated.

Again an analogue to Bayes' rule holds, now for the global variable for mean of the total number of pairs, but this time only as another approximation $\langle II\rangle_1 \approx \langle I|I_{\mathrm{nn}}=1\rangle \cdot \langle I\rangle$ with calculations parallel to those for the local analogue of Bayes' rule. It is

$$\langle II\rangle_1 := \sum_{i=1}^{N}\sum_{j=1}^{N} J_{ij} \underbrace{\langle I_i I_j\rangle}$$

$$= \langle I_i|I_j=1\rangle \cdot \langle I_i\rangle$$

exact analogue of Bayes' rule for local means

$$\approx \sum_{j=1}^{N} \underbrace{\left(\sum_{i=1}^{N} J_{ij}\langle I_i|I_j=1\rangle\right)}_{=:\langle I|I_{\mathrm{nn},j}=1\rangle} \cdot \langle I_j\rangle \quad . \tag{2.51}$$

mean conditioned on nearest neighbors

Hence

$$\langle II\rangle_1 = \sum_{j=1}^{N} \underbrace{\langle I|I_{\mathrm{nn},j}=1\rangle}_{\approx\langle I|I_{\mathrm{nn}}=1\rangle} \cdot \langle I_j\rangle$$

$$\approx \langle I|I_{\mathrm{nn}}=1\rangle \cdot \underbrace{\sum_{j=1}^{N}\langle I_j\rangle}_{\langle I\rangle}$$

$$= \langle I|I_{\mathrm{nn}}=1\rangle \cdot \langle I\rangle$$

such that in total we obtain

$$\langle II\rangle_1 \approx \langle I|I_{\mathrm{nn}}=1\rangle \cdot \langle I\rangle \tag{2.52}$$

or

$$\langle I|I_{\mathrm{nn}}=1\rangle \approx \frac{\langle II\rangle_1}{\langle I\rangle} \quad . \tag{2.53}$$

Inserting Eq. (2.53) into Eq. (2.44) finally gives the pair approximation for the total number of triples

$$\langle III\rangle_{1,1} \approx \frac{\langle II\rangle_1 \cdot \langle II\rangle_1}{\langle I\rangle} \quad . \tag{2.54}$$

To obtain the above pair approximation, the following two approximation steps have been used:
(i)

$$\langle I_i I_j I_k\rangle = \underbrace{\langle I_i|I_j=1, I_k=1\rangle}_{\approx\langle I_i|I_j=1\rangle} \cdot \langle I_i I_j\rangle \tag{2.55}$$

or in the mean $\langle I_i|I_j=1, I_k=1\rangle := \sum_{I_i=0}^{1} I_i p(I_i|I_j=1, I_k=1)$ as independence in the conditional probabilities $p(I_i|I_j=1, I_k=1) \approx p(I_i|I_j=1)$, and the second approximation:
(ii)

$$\langle I_i|I_{\mathrm{nn},j}=1\rangle := \sum_{i=1}^{N} J_{ij}\langle I_i|I_j=1\rangle \approx \langle I_i|I_{\mathrm{nn}}=1\rangle \tag{2.56}$$

independent of the site index j.

2.4 The SIS Dynamics under Pair Approximation

For the SIS-dynamics we have the ODE equations

$$\frac{d}{dt}\langle I\rangle = b\langle SI\rangle_1 - a\langle I\rangle \tag{2.57}$$

and with $\langle SI\rangle_1 = Q\langle I\rangle - \langle II\rangle_1$

$$\frac{d}{dt}\langle I\rangle = b\left(Q\langle I\rangle - \langle II\rangle_1\right) - a\langle I\rangle \tag{2.58}$$

while for the pairs

$$\frac{d}{dt}\langle II\rangle_1 = 2b\langle ISI\rangle_{1,1} - 2a\langle II\rangle_1 \quad . \tag{2.59}$$

For triples $\langle ISI\rangle$ as opposed to $\langle SIR\rangle$ we have to approximate the true triples $\widetilde{\langle ISI\rangle}$ defined as

$$\widetilde{\langle ISI\rangle}_{1,1}(t) := \sum_{i=1}^{N}\sum_{j=1}^{N}\sum_{k=1,k\neq i}^{N} J_{ij}J_{jk}\langle I_iS_jI_k\rangle \tag{2.60}$$

by

$$\widetilde{\langle ISI\rangle} \approx \frac{Q-1}{Q}\cdot\frac{\langle IS\rangle\cdot\langle SI\rangle}{\langle S\rangle} \quad . \tag{2.61}$$

Then all formal triples are given by

$$\langle ISI\rangle = \widetilde{\langle ISI\rangle} + \langle SI\rangle \quad . \tag{2.62}$$

As a pair approximation we can apply

$$\langle ISI\rangle_{1,1} \approx \frac{Q-1}{Q}\cdot\frac{(Q\langle I\rangle - \langle II\rangle_1)^2}{N-\langle I\rangle} + Q\langle I\rangle - \langle II\rangle_1 \quad . \tag{2.63}$$

Thus we obtain with $x := \langle I\rangle/N \in [0,1]$ and $y := \langle II\rangle_1/(NQ) \in [0,1]$ the closed ODE system

$$\frac{dx}{dt} = bQ(x-y) - ax \tag{2.64}$$

and

$$\frac{dy}{dt} = 2b(Q-1)\frac{(x-y)^2}{1-x} + 2b(x-y) - 2ay \tag{2.65}$$

with appropriate starting conditions x_0 and y_0, e.g. a fixed starting value for all runs in a simulation would correspond to $y_0 = 0$.

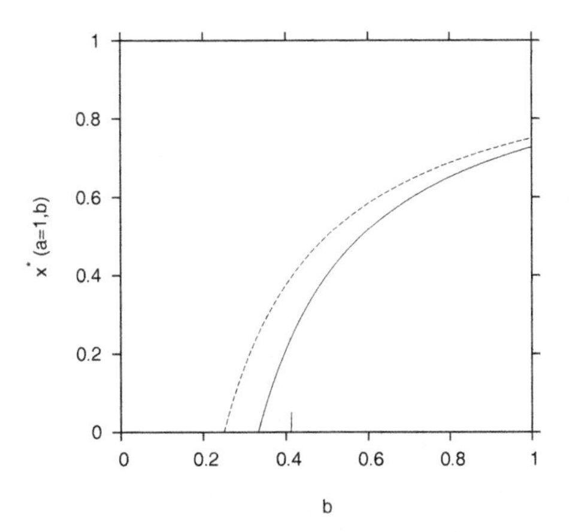

Fig. 2.4 For the SIS-system we compare the mean field solution (dashed line) with the pair approximation (full line) for $x^*(b)$ as obtained from Eq. (2.66). Furthermore, the critical value for the contact process is given as well, as obtained from extended spatial stochastic simulation reported in the literature as $\beta_C = 0.4122$. The pair approximation solution approaches the simulation value better than the mean field solution. (Parameters $Q = 4$ appropriate for two dimensional spatial systems and $a = 1$.)

The stationary states of the pair approximated birth and death process or SIS spatial epidemics can easily be calculated from Eqs. (2.64) and (2.65). One stationary state is again the extinction, i.e. $x_1^* = 0$ and $y_1^* = 0$. The non-trivial solution is given by

$$x_2^* = \frac{bQ(Q-1) - aQ}{bQ(Q-1) - a} \tag{2.66}$$

and

$$y_2^* = \frac{bQ - a}{bQ} \cdot x_2^* \tag{2.67}$$

where the first term of the product is the stationary state of the mean field approximation $\frac{bQ-a}{bQ} = \left(1 - \frac{a}{bQ}\right)$. Also the stationary state of the mean proportion of infected x_2^* approximates for large numbers of neighbors $Q \to \infty$ the mean field solution

$$x_2^* = \frac{bQ(Q-1) - aQ}{bQ(Q-1) - a} = \frac{b(Q-1) - a}{b(Q-1) - \frac{a}{Q}} \to \frac{bQ - a}{bQ} = \left(1 - \frac{a}{bQ}\right) \tag{2.68}$$

since then $(Q-1) \approx Q$ and $\frac{a}{Q} \to 0$. Figure 2.4 shows a comparison of stationary values of mean field approximation versus pair approximation and the numerical value for the true stochastic spatial system reported in the literature [Dickman and da Silva (1998)] as as $b_c = 0.4122$ (the value they [Dickman and da Silva (1998)] give is for their λ which is related to our variable as $b = \lambda/4$). The pair approximation solution is already half way better than the mean field solution.

The critical value for the stationary state in the pair approximation is obtained from $x_2^* = \frac{bQ(Q-1)-aQ}{bQ(Q-1)-a} = x_1^* = 0$, giving

$$b_c = \frac{a}{Q-1} \tag{2.69}$$

and for two spatial dimensions, hence $Q = 4$, and fixed $a := 1$ given by $b_c = \frac{a}{Q-1} = \frac{1}{3}$, whereas for mean field we obtain $b_c = \frac{a}{Q} = \frac{1}{4}$. Hence the pair approximation critical value is closer to the critical value $b_c = 0.4122$ for the explicit spatial system than for the mean field critical value.

2.5 Conclusions and Further Reading

Non-spatial stochastic processes, as described in, for example, [van Kampen (1992)] for chemical and physical processes, have been applied to biology for a long time [Goel and Richter-Dyn (1974)], whereas spatial aspects have more recently enjoyed increasing attention among biologists, especially ecologists and epidemiologists [Keeling, Rand and Morris (1997)]. For an overview of the development during the 1990s see [Rand (1999)], and more recently [Dieckmann, Law and Metz (2000)].

As a starting point we have used the master equation approach for a spatial system as for example used in [Glauber (1963)] and derive from it equations for the dynamics of moments, which under additional assumptions give closed ODE systems (moment closure methods). Such ODE systems have very recently been used to manage real world epidemics [Ferguson, Donnelly and Anderson (2001)]. In the easiest moment closure, which is the mean field assumption, the usual ODEs are found which were classically used as starting points for deterministic models. We have shown this explicitly for the easiest SIS-model. The approach can be applied easily to more complicated models with some more writing effort [Joo and Lebowitz (2004); Stollenwerk, Martins and Pinto (2007); Martins, Pinto and Stollenwerk (2009)].

Chapter 3

Criticality in Equilibrium Systems

As opposed to the birth–death process, as we have seen in Chapter 2, the so-called stochastic equilibrium systems have an interesting stationary state where most aspects of criticality can be studied and the notions then transfered to the non-equilibrium systems. The stochastic birth–death process has as its only stationary state the "empty state", also called the "absorbing state", whereas its mean field solution shows different stationary states and a transition between them. In contrast, equilibrium systems already show, in the analytic stationary state of the stochastic formulation, all characteristics of non-trivial stationary states and transitions between them, called phase transitions for historic reasons.

Since most of the notions of criticality have their origin in equilibrium systems, we will investigate in this chapter the most prominent example, the Ising model, and one of its dynamic extensions, the Glauber dynamics. The Glauber dynamics is formulated as a spatial master equation in complete analogy to the birth–death process of Chapter 2, and is also for a two-valued variable, in this case the spin variable. Only the transition rates are different, resulting in a non-trivial stationary state. Historically, the stationary state of spin systems was known long before Glauber, this is the Ising spin system, and Glauber constructed his dynamics in a way to obtain the Ising spin stationary state from his model.

3.1 The Glauber Model: Stochastic Dynamics for the Ising Model

The Glauber model is given as a spatial master equation, and is analogous to the spatial birth and death process, just for the two-valued variable $\sigma_i \in \{-1, +1\}$, the spins pointing up or down, instead of the variable

$I_i \in \{0, 1\}$, for occupied or non-occupied lattice sites in the birth–death process. The master equation of the Glauber model is

$$\frac{d}{dt} p(\sigma_1, ..., \sigma_N, t) = \sum_{i=1}^{N} w_{\sigma_i, -\sigma_i} \; p(\sigma_1, ..., -\sigma_i, ..., \sigma_N, t)$$
$$- \sum_{i=1}^{N} w_{-\sigma_i, \sigma_i} \; p(\sigma_1, ..., \sigma_i, ..., \sigma_N, t) \tag{3.1}$$

with transition rates $w_{\sigma_i, -\sigma_i}$ and $w_{-\sigma_i, \sigma_i}$ to be specified. The transition rates are specified such that the stationary state has the form

$$p^*(\sigma_1, ..., \sigma_N) = \frac{1}{Z} \cdot e^{\mathcal{H}(\sigma_1, ..., \sigma_N)} \tag{3.2}$$

with a function $\mathcal{H}$ of the form

$$\mathcal{H}(\sigma_1, ..., \sigma_N) = \sum_{i=1}^{N} \sum_{j=1}^{N} V_{ij} \sigma_i \sigma_j + \sum_{i=1}^{N} h_i \sigma_i + \sum_{i=1}^{N} c_i \tag{3.3}$$

and a normalization constant, the partition function Z. This stationary state Eq. (3.13) with Eq. (3.14) defines the Ising model for investigating a very simple thermodynamic model to understand spontaneous magnetization, as Ernst Ising first suggested in 1925 [Ising (1925)]. The parameters in the model are $V_{ij} := V \cdot J_{ij}$ with coupling strength V between neighboring spins and adjacency matrix $J_{ij} \in \{0, 1\}$ as already used for the birth and death process in Chapter 2, and h_i the external magnetic field acting on the spins. As a reasonable assumption, a constant field acting on all spins equally, $h_i =: h$, would be a second parameter in the model next to the coupling strength V.

In order to obtain reasonably sized numbers in numerical simulations of the Ising model, we use $C = \sum_{i=1}^{N} c_i = -N \ln 2$ for the 2^N possible configurations of N spins up and down, hence the partition function is

$$Z = \sum_{\sigma_1 = \pm 1} ... \sum_{\sigma_N = \pm 1} e^{\sum_{i=1}^{N} \sum_{j=1}^{N} V_{ij} \sigma_i \sigma_j + \sum_{i=1}^{N} h_i \sigma_i} \cdot \frac{1}{2^N} \quad . \tag{3.4}$$

The classical analysis of the Ising model considers changes of the distribution $p^*(\sigma_1, ..., \sigma_N) = (1/Z) e^{\mathcal{H}}$, or it considers global quantities like the magnetization $M := \sum_{i=1}^{N} \sigma_i$, by varying the model parameters V and h as shown in the next section. The mean number of infected $\langle I \rangle$ as a function of the rates a and b is the correspondence in the SIS-epidemic (respectively mean number of individuals in the birth and death process). In order to also investigate the time-dependent aspects of physical systems essentially described by the Ising model, one has to notate a time-dependent dynamic process, as Roy Glauber did in 1963 [Glauber (1963)]. The only requirement is to define the stochastic process in such a way to obtain as its stationary state the Ising model distribution. There are other possibilities of dynamics fulfillig the same stationary state, as e.g. the Kawasaki dynamics.

The transition rates of the master equation, Eq. (3.1), can be fixed in the following way. Stationarity means $dp^*(\sigma_1, ..., \sigma_N)/dt = 0$. Hence

$$\sum_{i=1}^{N} w_{\sigma_i,-\sigma_i} \; p^*(\sigma_1, ..., -\sigma_i, ..., \sigma_N) = \sum_{i=1}^{N} w_{-\sigma_i,\sigma_i} \; p^*(\sigma_1, ..., \sigma_i, ..., \sigma_N) \tag{3.5}$$

and when all terms under the sums are equal, which is called *detailed balance*, we obtain

$$\frac{w_{\sigma_i,-\sigma_i}}{w_{-\sigma_i,\sigma_i}} = \frac{p^*(\sigma_1, ..., \sigma_i, ..., \sigma_N)}{p^*(\sigma_1, ..., -\sigma_i, ..., \sigma_N)} \quad . \tag{3.6}$$

Detailed balance does not only make physical sense in equilibrium systems, but also guarantees the convergence of the stochastic dynamics to a single stationary state, see e.g. [Landau and Binder (2000)]. Of course, by construction, we will fix this single stationary state to be the distribution of the Ising model, hence

$$\frac{w_{\sigma_i,-\sigma_i}}{w_{-\sigma_i,\sigma_i}} = \frac{e^{\sigma_i\left(h_i+\sum_{j=1}^{N} V_{ij}\sigma_j\right)}}{e^{-\sigma_i\left(h_i+\sum_{j=1}^{N} V_{ij}\sigma_j\right)}} \quad . \tag{3.7}$$

With $e^{\sigma x}/\cosh(x) = 1 + \sigma \tanh(x)$ and $e^{-\sigma x}/\cosh(x) = 1 - \sigma \tanh(x)$ for a spin variable $\sigma \in \{-1, +1\}$ we can also obtain

$$\frac{w_{\sigma_i,-\sigma_i}}{w_{-\sigma_i,\sigma_i}} = \frac{1 + \sigma_i \tanh\left(h_i + \sum_{j=1}^{N} V_{ij}\sigma_j\right)}{1 - \sigma_i \tanh\left(h_i + \sum_{j=1}^{N} V_{ij}\sigma_j\right)} \quad . \tag{3.8}$$

So the transition rates of the master equation of the Glauber model can be given as

$$w_{\sigma_i,-\sigma_i} = \lambda \cdot e^{\sigma_i\left(h_i+\sum_{j=1}^{N} V_{ij}\sigma_j\right)} \quad , \tag{3.9}$$

and

$$w_{-\sigma_i,\sigma_i} = \lambda \cdot e^{-\sigma_i\left(h_i+\sum_{j=1}^{N} V_{ij}\sigma_j\right)} \quad , \tag{3.10}$$

or since a common factor does not alter the stationary state due to detailed balance

$$w_{\sigma_i,-\sigma_i} = \lambda \cdot \left(1 + \sigma_i \tanh\left(h_i + \sum_{j=1}^{N} V_{ij}\sigma_j\right)\right) \quad , \tag{3.11}$$

and

$$w_{-\sigma_i,\sigma_i} = \lambda \cdot \left(1 - \sigma_i \tanh\left(h_i + \sum_{j=1}^{N} V_{ij}\sigma_j\right)\right) \quad , \tag{3.12}$$

with λ a rate giving the transient time scale.

In one dimension, the transition rates can even be expressed explicitly as polynomials of the spin variable [Glauber (1963)], looking very similar to those for the birth and death process. However, due to the special features of the Glauber model, namely the detailed balance, it behaves qualitatively very differently to that used in the birth–death process. No absorbing state is present here.

3.1.1 *A first glance at the dynamic Ising model*

Fig. 3.1 The Glauber model, a stochastic simulation in a two-dimensional spin system near the phase transition.

A stochastic simulation of the Glauber model in two dimensions after a transient a lattice state is shown in Fig. 3.1, here close to the phase transition point. There are areas of negative spins (white) and positive (black) that are rather mixed but of very different sizes, so the system cannot decide if it becomes completely black (with some small white points inbetween) or completely white (with some small black points inbetween) as we would expect in the ordered phase, i.e. for large coupling strength V.

Furthermore, the picture does not only contain very fine grizzling between black and white, as we would see on a television set without program but only showing white noise, but larger areas of ordering. In the unordered phase, i.e. for small coupling strength V, we would expect such uncorrelated white noise,

3.2 The Ising Model, a Paradigm for Equilibrium Phase Transitions

Ernst Ising suggested a very simple thermodynamic model to understand spontaneous magnetization [Ising (1925)]. Statistical mechanics, as used by Ising, considers equilibrium distributions

$$p^*(\sigma_1, ..., \sigma_N) = \frac{1}{Z} \cdot e^{\mathcal{H}(\sigma_1, ..., \sigma_N)} \tag{3.13}$$

and quantities derived from them, like the magnetization per spin $\langle M \rangle := \langle \frac{1}{N} \sum_{i=1}^{N} \sigma_i \rangle$ and magnetic susceptibility $\chi := \langle M^2 \rangle - \langle M \rangle^2$. Here, the N spins $\{\sigma_i\}_{i=1}^N$, $\sigma_i \in \{-1, +1\}$, are distributed in $p^*(\sigma_1, ..., \sigma_N)$ according to Boltzmann weights $e^{\mathcal{H}}$ (originally introduced to represent well-known distributions in ideal gas theory, the Maxwell distribution of velocities) with a function

$$\mathcal{H}(\sigma_1, ..., \sigma_N) = \sum_{i=1}^{N} \sum_{j=1}^{N} V_{ij} \sigma_i \sigma_j + \sum_{i=1}^{N} h_i \sigma_i + \sum_{i=1}^{N} c_i \tag{3.14}$$

and a normalization, the partition function Z. In order to obtain reasonably-sized numbers we use $C = \sum_{i=1}^{N} c_i = -N \ln 2$ for the 2^N possible configurations of spins up and down, hence

$$Z = \sum_{\sigma_1 = \pm 1} ... \sum_{\sigma_N = \pm 1} e^{\sum_{i=1}^{N} \sum_{j=1}^{N} V_{ij} \sigma_i \sigma_j + \sum_{i=1}^{N} h_i \sigma_i} \cdot \frac{1}{2^N} \quad . \tag{3.15}$$

It is $V_{ij} := V \cdot J_{ij}$ with coupling strength V and adjacency matrix $J_{ij} \in \{0, 1\}$ and $h := h_i$ the external magnetic field.

Classically, the interest in thermodynamics focused on the description of macroscopic quantities, such as the total magnetization M and its dependence on the parameters external magnetic field h and coupling strength V (here standing for inverse temperature)

$$M = M(V, h) \quad . \tag{3.16}$$

Boltzmann introduced the modelling of such macroscopic quantities by underlying stochastic processes of the microvariables. In magnetic systems, these are the single spins σ_i. This origin in classical thermodynamics of macroscopic variables explains the form of $p^* = e^{\mathcal{H}}/Z$ and the function $\mathcal{H}(\sigma_1, ..., \sigma_N)$ as essentially the equi-distribution of microscopic variables, with derivation via the maximization of entropy.

The interest in phase transitions, especially the magnetic transition, led to the analysis of models like the Ising model, always assuming very large systems, N in the limit of infinity, the thermodynamic limit.

For finite, especially small, systems of stochastic processes, it is very difficult to properly define and measure the effects of phase transitions, since the well-known basic quantity to describe phase transitions in infinite systems, the Gibbs free energy [Zinn-Justin (1989)], performs poorly in small systems and shows some subjectivity in interpretation (flat versus dish-shaped behavior around phase transition points [Le Bellac (1991)]).

Instead of using the Gibbs free energy, we will investigate the distribution of the magnetization, a quantity which is difficult to observe in large (i.e. most physical) systems, but easily accessible in small systems (i.e. computational, and also population biological or epidemiological systems which are small compared to $N \approx 10^{23}$ in typical physical applications). The negative logarithm of the distribution is sometimes called an estimate of the behavior of the Gibbs free energy for system size going towards infinity.

We will try to retrieve all relevant quantities for describing a phase transition from our investigation of the magnetization distribution, in analogy to the description of infinite systems phase transitions on the basis of the Gibbs free energy. Finally, we aim to generalize to non-equilibrium phase transitions, especially directed percolation, which describes criticality in birth–death processes.

The distribution of the total magnetization is given by

$$p(M) = \sum_{\sigma_1=\pm 1} ... \sum_{\sigma_N=\pm 1} \delta\left(M - \sum_{i=1}^{N} \sigma_i\right) p^*(\sigma_1, ..., \sigma_N) \tag{3.17}$$

from simply applying Bayes' rule

$$p(M) = \sum_{\sigma_1=\pm1} ... \sum_{\sigma_N=\pm1} p(M|\sigma_1, ..., \sigma_N) \cdot p(\sigma_1, ..., \sigma_N) \tag{3.18}$$

for joint and conditional probabilities. While the total magnetization goes from $-N$ to N, and so can give quite big numbers for large physical systems, the magnetization per spin M/N is bound between -1 and 1, no matter how big the system is. For convenience, in the graphics we will often plot the magnetization per spin rather than the total magnetization. In the calculations, the constant N does not give any qualitative difference.

The mean magnetization, defined as $\langle M \rangle := \sum_{i=0}^{N} M \cdot p(M)$ with variable $M/N = -1 + i \cdot \Delta M$ and $\Delta M = 2/N$, is given by

$$\langle M \rangle = \sum_{\sigma_1=\pm1} ... \sum_{\sigma_N=\pm1} \left(\sum_{i=0}^{N} M \cdot \delta \left(M - \sum_{j=1}^{N} \sigma_j \right) \right) p^*(\sigma_1, ..., \sigma_N) \quad . \tag{3.19}$$

Using the δ-function, the mean magnetization is then

$$\langle M \rangle = \sum_{\sigma_1=\pm1} ... \sum_{\sigma_N=\pm1} \left(\sum_{i=1}^{N} \sigma_i \right) \frac{1}{Z} e^{\mathcal{H}} \quad . \tag{3.20}$$

The mean magnetization $\langle M \rangle$ is an easily accessible quantity in physical systems, whereas the distribution of the magentization $p(M)$ is easily accessible in small systems. Furthermore, the mean $\langle M \rangle$ is dominated by the maximum M_{max} of the distribution $p(M)$ for large systems. Hence the mean magnetization

$$\langle M \rangle = \langle M \rangle(V, h) \tag{3.21}$$

becomes a function of the model parameters V and h.

In the dynamic Ising model, the Glauber model, the mean magnetization becomes a time-dependent function, in analogy to the total number of individuals in the birth–death process in Chapter 2

$$\langle M \rangle(t) = \sum_{\sigma_1=\pm1} ... \sum_{\sigma_N=\pm1} \left(\sum_{i=1}^{N} \sigma_i \right) p(\sigma_1, ..., \sigma_N, t) \tag{3.22}$$

where the master equation of that Glauber model determines the dynamics of the distribution $p(\sigma_1, ..., \sigma_N, t)$.

In complete analogy, in epidemiology we have seen the local variable I_i as a two-valued variable and its global counterpart $I = \sum_{i=1}^{N} I_i$. By defining

$$I_i := (\sigma_i + 1)/2 \tag{3.23}$$

the variable $\sigma_i \in \{-1, +1\}$ can be transformed into the variable $I_i \in \{0, 1\}$ and vice versa $\sigma_i = 2 \cdot I_i - 1$. Interestingly, in physics, models for the variable $I_i \in \{0, 1\}$ are also considered. They describe a gas moving around in spaces (normally on a simple square lattice), the so-called lattice gas, in the same way as we discussed for the birth–death process. Again, for epidemics the distribution of

$$p(I) = \sum_{I_1=0}^{1} \dots \sum_{I_N=0}^{1} \delta\left(I - \sum_{i=1}^{N} I_i\right) p(I_1, ..., I_N) \tag{3.24}$$

can be studied with the total number of infected $I \in \{0, 1, 2, ..., N\}$. The relation between total number and "magnetization" per site would be given by $M = \frac{2 \cdot I}{N} - 1$. The mean total number again is given by

$$\langle I \rangle = \sum_{I_1=0}^{1} \dots \sum_{I_N=0}^{1} \left(\sum_{i=1}^{N} I_i\right) p(I_1, ..., I_N) \quad . \tag{3.25}$$

as seen Chapter 2. The analogy between the formulation of the magnetic system and the biological becomes evident with these transformations.

3.2.1 *Distribution of magnetization and Gibbs free energy*

Here we give a brief overview of the classical thermodynamic quantities used to describe equilibrium phase transitions. This might help the reader from other fields to understand the language sometimes used to also describe non-equilibrium phase transitions. Zinn-Justin's comments on the overlap in language between two fields, in his case quantum field theory and critical phenomena [Zinn-Justin (1989)], and hence cross-fertilization, also holds for equilibrium phase transitions and non-equilibrium phase transitions.

The distribution of the total magnetization $p(M)$ can be interpreted as an estimate for the Gibbs free energy. This is an analogue to the Helmholtz free energy $F(V, h) := -\ln Z$, but gives thermodynamic quantities in different natural variables. The estimate is

$$p(M) \sim e^{-\Gamma_V(M,h)} \tag{3.26}$$

with convergence for large systems $N \to \infty$ [Binney et al. (1992)]. Classical thermodynamics starts from there by defining

$$\Gamma_V(M,h) := F(V,h) + Mh \tag{3.27}$$

and

$$\frac{\partial F}{\partial h} = \frac{\partial}{\partial h}(-\ln Z) \tag{3.28}$$

then

$$\frac{\partial}{\partial h}(-\ln Z) = -\frac{1}{Z}\sum_{\sigma_1=\pm 1} \cdots \sum_{\sigma_N=\pm 1}\left(\sum_{i=1}^{N}\sigma_i\right) e^{V\sum_{i=1}^{N}\sum_{j=1}^{N}\sigma_i\sigma_j + h\sum_{i=1}^{N}\sigma_i + C} \tag{3.29}$$

hence

$$\frac{\partial}{\partial h}(-\ln Z) = -\langle\sum_{i=1}^{N}\sigma_i\rangle =: -M \quad . \tag{3.30}$$

In the last step we simply defined $M := \langle\sum_{i=1}^{N}\sigma_i\rangle$, which would be the mean of the total magnetization $\langle M\rangle$. Since the thermodynamic relations between F, h and M historically preceed the statistical interpretation, the mean brackets are often omitted.

So we have $\partial F/\partial h = -M$. Here we have used the special form of the Boltzmann distribution $p^* = (1/Z)e^{\mathcal{H}}$ to obtain $\partial F/\partial h = -\langle\sum_{i=1}^{N}\sigma_i\rangle$. From this we obtain the famous thermodynamic relation for the total derivative of F

$$dF := \frac{\partial F}{\partial h}\, dh + \frac{\partial F}{\partial V}\, dV = -M\, dh + \frac{\partial F}{\partial V}\, dV \tag{3.31}$$

and the Legendre transformed of the Helmholtz free energy F defines the Gibbs free energy $\Gamma_V(M,h) := F(V,h) + Mh$ as

$$d\Gamma := dF + M\, dh + h\, dM = \frac{\partial F}{\partial V}\, dV + h\, dM \quad . \tag{3.32}$$

The second moment of the magnetization has a similar relation to the thermodynamic potentials, here F. It is

$$\langle\sum_{i=1}^{N}\sum_{i=j}^{N}\sigma_i\sigma_j\rangle = -\frac{\partial F}{\partial V} \sim \chi \quad . \tag{3.33}$$

The magnetic susceptibility gives an expression for the remaining term $\frac{\partial F}{\partial V}\, dV$. However, then the exact definition for the magnetic susceptibility is $\chi = \langle \sum_{i=1}^{N} \sum_{i=j}^{N} \sigma_i \sigma_j \rangle - \langle \sum_{i=1}^{N} \sigma_i \rangle^2 = (\partial^2/\partial h^2)\ln Z$ as it will be used below.

3.3 Equilibrium Distribution around Criticality

3.3.1 *Distribution of magnetization*

We will now have a closer look at the distribution of the total magnetization

$$p(M) = \sum_{\sigma_1=\pm 1} \dots \sum_{\sigma_N=\pm 1} \delta \left(M - \sum_{i=1}^{N} \sigma_i \right) \frac{1}{Z} e^{\mathcal{H}_{V,h}(\sigma_1,\dots,\sigma_N)} \tag{3.34}$$

and its dependence on the model parameters V and h. Therefore, we perform a so-called complete enumeration of the system. This means that we calculate all possible configurations for the N spin variables $(\sigma_1, ..., \sigma_N)$, i.e. $(\sigma_1, ..., \sigma_N) = (1, 1, ..., 1, 1)$, then $(\sigma_1, ..., \sigma_N) = (1, 1, ..., 1, -1)$ etc. There are 2^N such possible configurations. For each configuration we calculate the Boltzmann weight

$$p^*(\sigma_1, ..., \sigma_N) = \frac{1}{Z} e^{\mathcal{H}_{V,h}(\sigma_1,\dots,\sigma_N)} \tag{3.35}$$

and its total magnetization

$$M(\sigma_1, ..., \sigma_N) = \sum_{i=1}^{N} \sigma_i \quad . \tag{3.36}$$

For each possible M value we then add up the Boltzmann weights of all configurations $(\sigma_1, ..., \sigma_N)$ which have this magnetization $M(\sigma_1, ..., \sigma_N)$. This last step is essentially what the delta-function in Eq. (3.34) does.

Thus, for each parameter set V and h, we have to calculate these 2^N configurations and its weight and magnetization. However, a modern computer is still only able to perform such complete enumeration for very small system sizes N; essentially too small to give good quantitative results for thermodynamic quantities such as the critical temperature (or critical coupling strength) and further quantities around the critical region of the Ising model. Therefore one still has to perform stochastic simulations with algorithms similar to the Glauber dynamics to obtain quantitative results. In turn, these results face the expected problems of stochastic simulations, the very long runs required for convergence and good estimates etc.

The complete enumeration of small systems gives a good qualitative insight into the behavior of the Ising model. Here we first perform a simulation showing a 4×4 system, hence $N = 16$, which is managable on a laptop in reasonable time, for the vanishing outer magnetic field $h = 0$ and three values for V. The results are shown in Fig. 3.2 for coupling strength close to the critical for this small system.

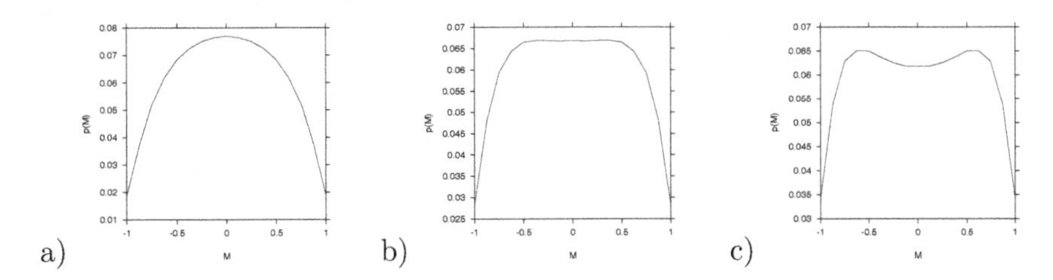

Fig. 3.2 Distribution $p(M)$ of the magnetization M in the 4×4 lattice in two dimensions for coupling V with a) $V = 0.13$, b) $V = 0.14 \approx V_c$, and c) $V = 0.145$. The external magnetic field is absent, $h = 0$.

In Fig. 3.2 for three values of the coupling constant V the distribution $p(M)$ of the magnetization is shown, each with absent external magnetic field, $h = 0$. In Fig. 3.2a) the system is subcritical, only one maximum at $M = 0$ appears. In Fig. 3.2c) two maxima $M_{max} \neq 0$ are found symmetrically around $M = 0$, where there is a local minumum in $p(M)$. For the value of $V = 0.14 \approx V_c$ we find one maximum at $M = 0$, but being broadened up just before splitting into the two symmetric maxima $M_{max} \neq 0$. This is the point of phase transition. Fluctuations can easily happen for values between $M \approx -0.5$ and $M \approx +0.5$, for which all magnetizations are roughly equally likely.

3.3.2 *External magnetic field*

Next, we simulate the toy system of 4×4 Ising spins below the critical coupling strength with a small negative and a small positive magnetic field h, compared to the simulation without the magnetic field (see Fig. 3.3).

The maximum and hence also the mean of the distribution $p(M)$ for the total magnetization M only change slightly. The system is not very sensitive to small changes of the external magnetic field h. Also for larger system sizes one would expect the peak of the distribution $p(M)$ to be sharper, hence the maximum essentially determines the behavior of the mean.

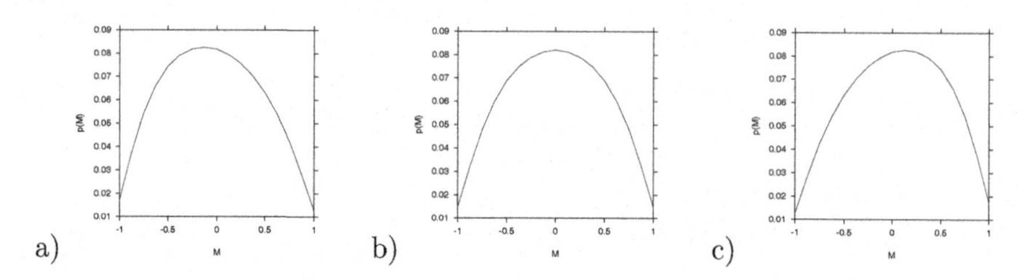

Fig. 3.3 Distribution $p(M)$ of the magnetization M in the 4×4 lattice in two dimensions for coupling $V = 0.125 << V_c$ with a) $h = -0.01$, b) $h = 0$, and c) $h = +0.01$.

However, when we simulate this system above the critical coupling strength (see Fig. 3.4), the picture changes dramatically. Due to the symmetric hump around zero, the mean value for vanishing external field $h = 0$, Fig. 3.4b) is zero, since the masses of the two humps cancel each other out. But for small negative or small positive external field h, the hump on the negative side of the magnetization M or the one on the positive side dominates the distribution, and hence also the mean value of $p(M)$. Again, a larger system size would further enhance the effect.

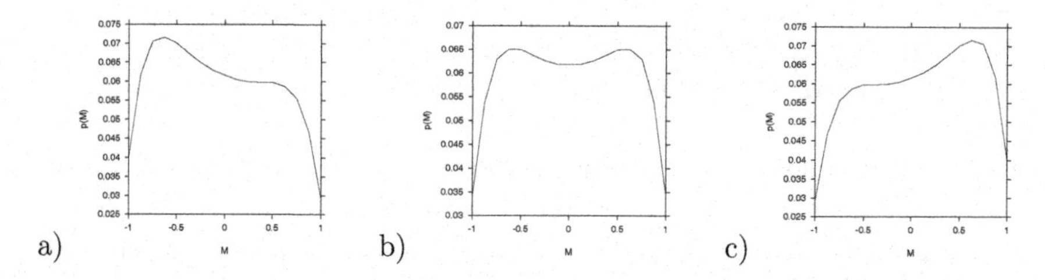

Fig. 3.4 Distribution $p(M)$ of the magnetization M in the 4×4 lattice in two dimensions for coupling $V = 0.145 > V_c$ with a) $h = -0.01$, b) $h = 0$, and c) $h = +0.01$.

To describe the phenomenon of a phase transition further, we will look at the behavior of the maximum of the distribution $p(M)$ as a substitute for the mean $\langle M \rangle$, which is classically used to describe the behavior of the Ising model around criticality. We should always bear in mind that for larger system size, the mean of the total magnetization will increasingly be determined by the maximum of the distribution. Surprisingly, from our toy system of only $N = 16$ spins we already find qualitatively the right behavior as investigated in extenso for larger systems with considerably more computational effort.

3.3.3 *The maximum of the total magnetization distribution*

Our small toy system has relatively few possible magnetizations, hence the resolution of the distribution is rather low in spite of the high computational effort of evaluating 2^N configurations. However, the observation that the distribution $p(M)$ (respectively its logarithm) can be rather well described by a fourth order polynomial can help in this respect.

3.3.3.1 *Approximation with Lagrange polynomials*

To approximate the logarithm of the distribution $p(M)$ by a polynomial $\pi(M)$, we use the Lagrange polynomials just as an interpolation between points M_ν of the original distribution $p(M)$, from the complete enumeration. Hence we have

$$\ln(p(M)) \approx \pi(M) := \sum_{\nu=1}^{n} \left(\ln(p(M_\nu)) \cdot \prod_{\varphi=1,\varphi\neq\nu}^{n} \frac{M - M_\varphi}{M_\nu - M_\varphi} \right) \tag{3.37}$$

for the 4th order Lagrange polynomial interpolation with support points $M_\nu \in \{-1, -1/2, 0, 1/2, 1\}$ and where we take the obtained poynomial as a rough description of the negative of the Gibbs free energy $\pi(M) \sim -\Gamma(M)$ as parameters change.

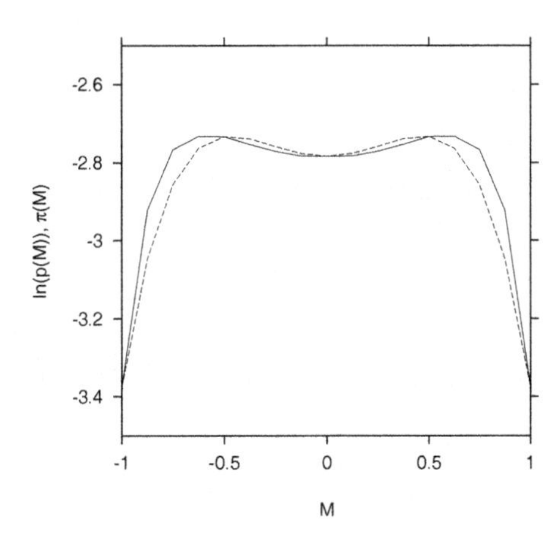

Fig. 3.5 Interpolation of five points of $p(M)$ (from full line) with a Lagrange polynomial of 4th order (dashed line). $V = 0.145$ and $h = 0$.

We also could use higher polynomial approximations. For the 6th order polynomial interpolation we use supporting points $M_\nu \in \{-1, -1/2, -\Delta M, 0, \Delta M, 1/2, 1\}$. However, the quality of the approximation improves little and becomes rather more complicated to interpret (see Fig. 3.6).

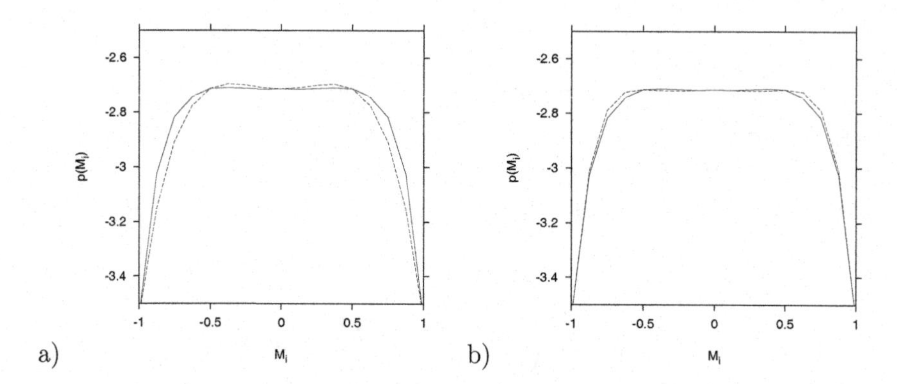

Fig. 3.6 a) Lagrange polynomial of 4th order, b) Lagrange polynomial of 6th order. $V = 0.140$ and $h = 0$. In the 6th order approximation, around the plateau there appear local maxima and minima which have no correspondence in the real physical system.

When comparing the approximations for 4th order (Fig. 3.6a) with the ones for 6th order Lagrange polynomials (Fig. 3.6b) the quality does not improve much, but new qualitative behavior shows up in the polynomial approximation which is not observed in the original Ising model distribution obtained by complete enumeration. Thus we can safely stick to the 4th order approximation.

3.3.3.2 *The maximum magnetization with changing parameters*

We will now use the approximation of the distribution of magnetization by the 4th order Lagrange polynomial under varying model parameters V and h. By using the polynomial approximation, we obtain information on the whole shape of $p(M)$ for evaluating the maximum. In contrast, if we only used the direct maximum of $p(M)$ from the complete enumeration of our toy Ising system, would not change gradually but in jumps of the discretization of $p(M)$. This way, we obtain some smoothing of the curves shown below, but qualitatively keep the essential information from the original spin system.

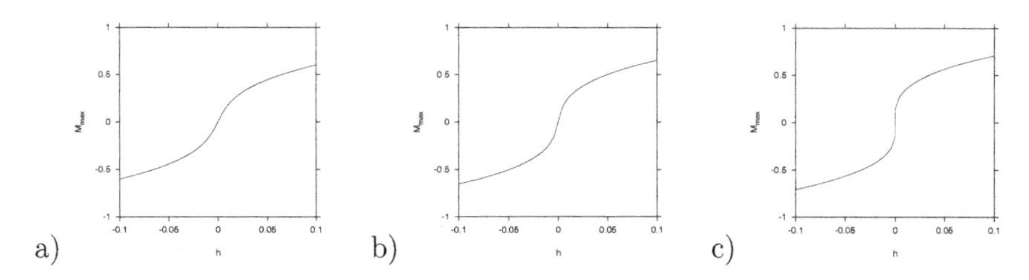

Fig. 3.7 Maximum of the polynomial $\pi(M) \approx \ln(p(M))$ as a function of the external magnetic field h for a) $V = 0.125$, b) $V = 0.130$ and c) $V = 0.135$.

In Fig. 3.7 we show with varying external magnetic field the change of the maximum of $p(M)$ and its approximation for various values of the coupling strength V. Figure 3.7a) for small coupling strength $V = 0.125$ the maximum of $p(M)$ changes very smoothly as the external magnetic field is changed from negative to positive values, with a finite slope crossing zero for both the maximum and the external field. In Fig. 3.7b) for a larger coupling strength $V = 0.13$, the curve is still smooth, but the slope around zero for both values becomes steeper, and even shows the first sign of a jump in Fig. 3.7c) for coupling strength $V = 0.135$.

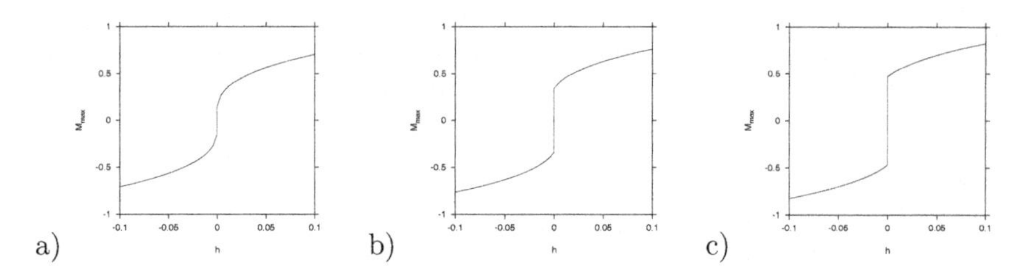

Fig. 3.8 Maximum M_{max} of $\pi(M) \approx \ln(p(M))$ as a function of the external magnetic field h for a) $V = 0.135$, b) $V = 0.140$ and c) $V = 0.145$.

Increasing the coupling strength further and further, as shown in Figs. 3.8a) to 3.8c), demonstrates the increasing jump height at zero external magnetic field. Interpreting these curves, we can conclude that the phase transition in the 4th order approximation of $p(M)$ occurs around the critical value for coupling strength of $V_c \approx 0.135$, whereas from inspection of the original curve for $p(M)$ the critical value seems to lie around 0.145. The difference is, of course, unsurprising due to the crude approximation with polynomials, but the order of magnitude is correct and the qualitative behavior around a phase transition can be demonstrated well for the various quantities.

Looking finally at the behavior of the maximum of the total magnetization at vanishing external field $h = 0$, we find the typical shape of a threshold behavior (see Fig. 3.9). In the graphics we only look for the first appearing maximum coming from the positive side in total magnetization.

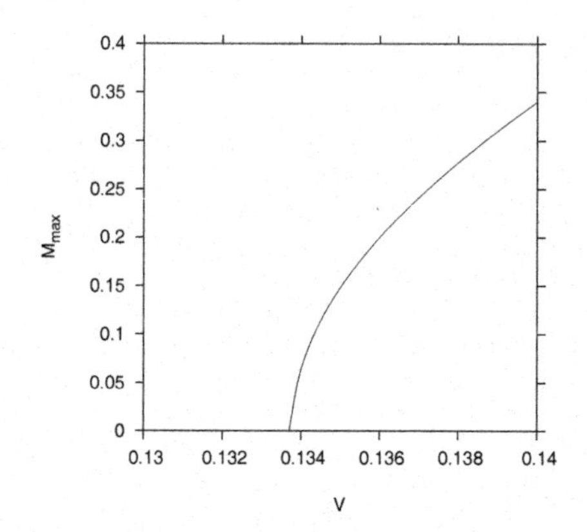

Fig. 3.9 For $h = 0$ and varying V the maximum M_{max} is shown around the critical value for the polynomial model (4th order) $V_{c4^{th}} = 0.13379$.

For small coupling strength V we observe zero magnetization as the maximal value for $p(M)$, whereas above a threshold value V_c the maximum lies well above zero. Figure 3.9 is comparable to the threshold behavior in birth and death processes (see Fig. 1.4a).

However, we cannot expect to obtain further quantitative results with the crude method of approximating the distribution $p(M)$ by a 4th order Lagrange polynomial, even while remembering that we are looking at a small system size. Instead, we will now look closer at the mean field approximation of the Ising spin system, to recover some of the above-seen qualitative features in an analytical way.

3.4 Mean Field Theory and its Exponents

In analogy to the mean field approximation for the birth–death process in Chapter 2, we will now apply the mean field approximation to the Ising model, i.e. the stationary state equation of the spin system. Unlike in the birth and death process, the dynamics are not approximated to obtain a closed ODE system. Instead, the mean magnetization will be expressed in a closed form by the approximation, giving a self-consistency equation. This self-consistency equation can be understood as the closing of the ODE system of the dynamics of the total number of individuals in the birth–death process considered earlier.

From the self-consistency equation the critical transition can be characterized, thus obtaining power laws. This is appealing as it can be used in non-equilibrium systems as well. We will obtain the mean field critical exponents of the Ising model, in order to be able to later use the concepts in the non-equilibrium systems fruitfully.

3.4.1 *Mean field self-consistency equation*

By inserting the mean field assumption into the partition function and evaluating the mean magnetization, we get an equation for the magnetization as function which again contains the magnetization. This defines a self-consistency equation that is only fulfiled for the correct mean field magnetization.

We use the mean field approximation as in Eq. (2.33) for the spin variable σ_i

$$\sum_{i=1}^{N}\sum_{j=1}^{N} V_{ij}\sigma_i\sigma_j = 2Q\langle\sigma\rangle\left(\sum_{i=1}^{N}\sigma_i\right) - Q\langle\sigma\rangle^2 \tag{3.38}$$

from neglecting the local variance $(\sigma_i - \langle\sigma\rangle)(\sigma_j - \langle\sigma\rangle)$. Furthermore,

$$m := \langle\sigma\rangle = \frac{M}{N} \tag{3.39}$$

is the magnetization per spin, whereas M is the total magnetization of the whole system of N spins. Hence, the mean magnetization is

$$\langle\sigma\rangle = \sum_{\sigma_1=\pm 1} \cdots \sum_{\sigma_N=\pm 1} \left(\frac{1}{N}\sum_{i=1}^{N}\sigma_i\right) p^*(\sigma_1, ..., \sigma_N) \tag{3.40}$$

with Eq. (3.13) $p^*(\sigma_1, ..., \sigma_N) = \frac{1}{Z} \cdot e^{\mathcal{H}(\sigma_1,...,\sigma_N)}$ and from Eq. (3.14)

$$\mathcal{H}(\sigma_1, ..., \sigma_N) = \sum_{i=1}^{N}\sum_{j=1}^{N} V_{ij}\sigma_i\sigma_j + \sum_{i=1}^{N} h_i\sigma_i + \sum_{i=1}^{N} c_i \tag{3.41}$$

and

$$\mathcal{H}(\sigma_1, ..., \sigma_N) \approx V\left(2Q\langle\sigma\rangle\left(\sum_{i=1}^{N}\sigma_i\right) - Q\langle\sigma\rangle^2\right) + h\left(\sum_{i=1}^{N}\sigma_i\right) + C =: \mathcal{H}_{m.f.} \tag{3.42}$$

using the mean field approximation Eq. (3.38), and $V_{ij} = V \cdot J_{ij}$. Also the partition function Eq. (3.15) has to be approximated in mean field, only using $\mathcal{H}_{m.f.}$

$$Z_{m.f.} = \sum_{\sigma_1=\pm1} \cdots \sum_{\sigma_N=\pm1} e^{(2QV\langle\sigma\rangle+h)\cdot(\sum_{i=1}^{N}\sigma_i)+(C-QV\langle\sigma\rangle^2)} \quad . \tag{3.43}$$

Hence, in the mean field approximation the magnetization is given by an expression again containing the magnetization

$$\begin{aligned}\langle\sigma\rangle &= \sum_{\sigma_1=\pm1} \cdots \sum_{\sigma_N=\pm1} \left(\frac{1}{N}\sum_{i=1}^{N}\sigma_i\right) \frac{1}{Z_{m.f.}} \cdot e^{\mathcal{H}_{m.f.}(\sigma_1,\ldots,\sigma_N)} \\ &= \frac{\sum_{\sigma_1=\pm1} \cdots \sum_{\sigma_N=\pm1} \left(\frac{1}{N}\sum_{i=1}^{N}\sigma_i\right) e^{\mathcal{H}_{m.f.}(\sigma_1,\ldots,\sigma_N)}}{\sum_{\sigma_1=\pm1} \cdots \sum_{\sigma_N=\pm1} e^{\mathcal{H}_{m.f.}(\sigma_1,\ldots,\sigma_N)}} \\ &= \frac{\frac{1}{N}\sum_{i=1}^{N}\left(\sum_{\sigma_i=\pm1}\sigma_i e^{\sigma_i(2QV\langle\sigma\rangle+h)} \sum_{\sigma_1=\pm1}\cdots\not\sum_{\sigma_i}\cdots\sum_{\sigma_N=\pm1} e^{(2QV\langle\sigma\rangle+h)\cdot(\sum_{i\neq j}\sigma_j)}\right)}{\sum_{\sigma_1=\pm1} \cdots \sum_{\sigma_N=\pm1} e^{(2QV\langle\sigma\rangle+h)\cdot(\sum_{i=1}^{N}\sigma_i)}}\end{aligned} \tag{3.44}$$

where the final step cannot be performed without help of the the mean field assumption. Hence

$$\begin{aligned}\langle\sigma\rangle &= \frac{1}{N}\sum_{i=1}^{N} \frac{\sum_{\sigma_i=\pm1}\sigma_i e^{\sigma_i(2QV\langle\sigma\rangle+h)}}{\sum_{\sigma_i=\pm1} e^{\sigma_i(2QV\langle\sigma\rangle+h)}} \\ &= \frac{1}{N}\sum_{i=1}^{N} \underbrace{\frac{e^{(2QV\langle\sigma\rangle+h)} - e^{-(2QV\langle\sigma\rangle+h)}}{e^{(2QV\langle\sigma\rangle+h)} + e^{-(2QV\langle\sigma\rangle+h)}}}_{\text{independent of the index } i} \\ &= \tanh(2QV\langle\sigma\rangle + h) \cdot \underbrace{\frac{1}{N}\sum_{i=1}^{N} 1}_{=1} \\ &= \tanh(2QV\langle\sigma\rangle + h)\end{aligned}$$

such that in total

$$\langle \sigma \rangle = \tanh(2QV\langle \sigma \rangle + h) \tag{3.45}$$

and resolving for h

$$h = \text{artanh}(m) - 2 \cdot QVm \tag{3.46}$$

with $m := \langle \sigma \rangle$. This is the so-called mean field self-consistency equation.

In Fig. 3.10 we plot for one value of coupling strength $V = V_c + \varepsilon$ with the critical value $V_c = 1/(2Q)$ and $\varepsilon = 0.02$ the mean field self-consistency equation, using the simple formula for the artanh-function

$$\text{artanh}(m) = \frac{1}{2}\log\left(\frac{1+m}{1-m}\right) \quad . \tag{3.47}$$

We plot for $m \in [-0.8, 0.8]$ and only invert graphically, i.e. calculating $h(m)$ and plotting m on the y-axis and h on the x-axis.

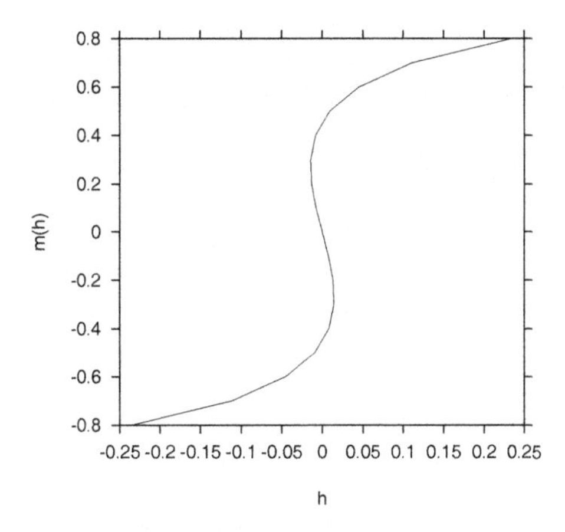

Fig. 3.10 Plot of the magnetization per spin of the Ising spin system as function of the external magnetic field with a coupling strength V slightly above its critical value, as obtained from the mean field self-consistency equation.

This Fig. 3.10 for the mean field equation can be compared to Fig. 3.8c) where the toy Ising system was evaluated by complete enumeration. The only difference here is that when the external magnetic field is zero, the original system (Fig. 3.8c) has infinite slope, whereas the mean field solution shows a hysteresis curve giving three values for vanishing external field. The infinite slope line in the original system connects the lowest with the highest point, hence giving a unique value for each value of the external field as can be expected in a physical system.

However, the hysteresis curve contains some information from the original physical system in the sense that in a stochastic simulation of the Ising model, e.g. in a Glauber dynamics, the system can stay for a long time in one of the two maxima of $p(M)$ when this maximum is already lower than that on the other side in the two-humped case, and only change eventually to the other maximum, i.e. here the other branch of the magnetization curve changing with external field.

We will now consider how several macroscopic quantities are derived from the mean field self-consistency equation and its behavior near the critical threshold of $V_c = 1/(2Q)$.

3.4.2 *Mean field quantities obtained from the self-consistency equation*

3.4.2.1 *Critical coupling strength in mean field approximation*

First, we must calculate the critical value of the coupling strength V. As seen above, the critical value V_c is determined by the transition from a one-humped to a two-humped distribution $p(M)$. Now, in mean field approximation we do not have easy access to a quantity like $p(M)$. However, as we have seen, the negative Gibbs free energy has the same qualitative behavior as the distribution of the magnetization (or at least the mean field solution of the Gibbs free energy, for a more detailed discussion see [Zinn-Justin (1989)]. The Gibbs free energy for spatial physical systems becomes flat above the phase transition and not two-humped).

Defining the Gibbs free energy

$$\Gamma := F + \sum_{i=1}^{N} M_i H_i = -\ln(Z) + N \cdot m \cdot h \tag{3.48}$$

and here in mean field approximation specifically for zero external magnetic field

$$\Gamma = -m^2 NVQ + N \cdot \frac{1}{2} \Big((1+m)\ln(1+m) + (1-m)\ln(1-m) \Big) \tag{3.49}$$

gives a criterium for determining the critical value V_c of the coupling strength.

At zero magnetization $m = 0$, the Gibbs free energy $\Gamma(m)$ changes from a local minimum to a local maximum, hence the second derivative changes

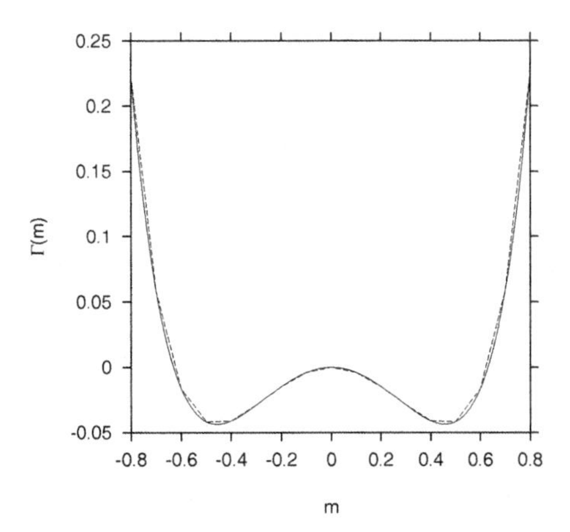

Fig. 3.11 The Gibbs free energy $\Gamma(m)$ in mean field approximation for coupling strength above the critical value, $V > V_c$.

its sign. So at the critical value V_c the second derivative of the Gibbs free energy at the point $m = 0$ is zero. The equation

$$\left.\frac{d^2\Gamma}{dm^2}\right|_{m=0} = 0 \tag{3.50}$$

gives the value for the critical coupling strength V_c, in detail

$$\frac{d^2\Gamma}{dm^2} = -2NVQ + \frac{2}{N}\left(\frac{1}{1+m} + \frac{1}{1-m}\right) \tag{3.51}$$

with $1/(1+m) + 1/(1-m) = 2$. Hence setting $\frac{d^2\Gamma}{dm^2} = 0$ we have

$$0 = -2NV_cQ + \frac{2}{N} \cdot 2 \tag{3.52}$$

giving

$$V_c = \frac{1}{2Q} \tag{3.53}$$

as critical value of the coupling strength in mean field approximation. In Fig. 3.11 the Gibbs free energy is plotted just above the critical coupling strength, showing in mean field approximation the typical two local minima shape.

3.4.2.2 *The first critical exponent in mean field approximation*

The first critical exponent β is given by the magnetization for a vanishing or infinitely small external magnetic field $h \to 0$ for large coupling $V > V_c$. From Eq. (3.46) in Taylor's expansion of artanh(m)

$$0 \stackrel{!}{=} h = \operatorname{artanh}(m) - 2 \cdot QVm \approx m + \frac{1}{3}m^3 + \frac{1}{5}m^5 ... - 2 \cdot QVm \quad (3.54)$$

we obtain with $V_c = 1/(2Q)$ and the higher orders in Taylor's expansion neglected, $(1/5)m^5 \to 0$,

$$0 \approx 2Q(V_c - V)m + \frac{1}{3}m^3 \quad (3.55)$$

hence

$$m \approx \sqrt{6Q}(V_c - V)^{\frac{1}{2}} \quad (3.56)$$

such that

$$m \sim (V_c - V)^{\beta} \quad (3.57)$$

with critical exponent $\beta = 1/2$ in mean field approximation.

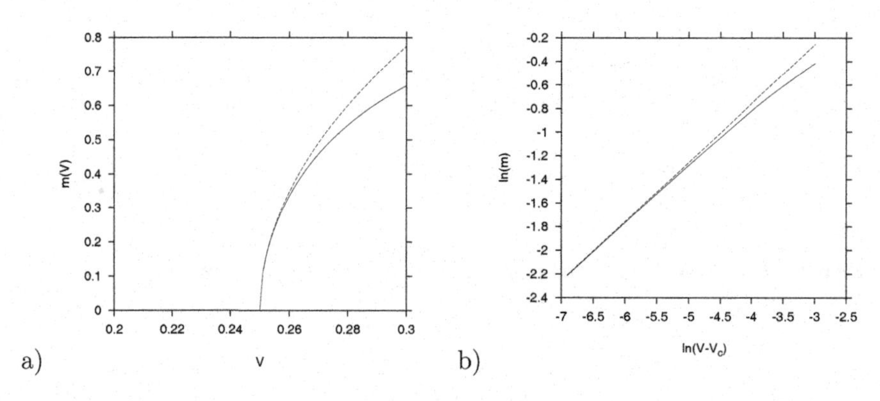

Fig. 3.12 a) $m(V)$ and b) $\log(m(\log(V - V_c)))$. In both cases the dashed curve is the approximation after truncating the power series, hence $m \approx \sqrt{6Q}(V_c - V)^{\frac{1}{2}}$, $\log(m) \approx \log(V_c - V) + \frac{1}{2}\log(6Q)$ respectively, and the lines are the zeros from $h = \operatorname{artanh}(m) - 2 \cdot QVm$ via Newton's method.

In Fig. 3.12a) the upper curve is obtained by $m = \sqrt{6Q}(V_c - V)^{\frac{1}{2}}$ from Eq. (3.56), whereas the lower curve is obtained from the mean field self-consistency equation Eq. (3.46) $h = \operatorname{artanh}(m) - 2 \cdot QVm$ via Newton's method to determine the zeros from $h(m) = 0$. The same for Fig. 3.12b) in double logarithmic scale immediately gives the critical mean field exponent $\beta = \frac{1}{2}$ as slope of the curves.

3.4.3 *Universal scaling function and exponent δ in mean field*

The equation of state, Eq. (3.46), also gives a universal scaling function $\varphi(x) = (1/3) - 2Q \cdot (1/x^2)$, once the exponent β is known. Again we look at $V > V_c$. Since $m \sim (V - V_c)^{\frac{1}{2}}$ we reorganize

$$h = \operatorname{artanh}(m) - 2 \cdot QVm \approx -2Q(V - V_c)m + \frac{1}{3}m^3 + \frac{1}{5}m^5 + ... \quad (3.58)$$

in a way containing the term $m/(V - V_c)^\beta$ which remains constant for $m \to 0$, whereas higher orders in m go to zero,

$$h \approx m^3 \left(\frac{1}{3} - 2Q \cdot \left(\frac{m}{(V - V_c)^{\frac{1}{2}}} + \frac{1}{5}m^2 + ... \right)^{-2} \right) \quad (3.59)$$

giving the scaling form

$$h \approx m^\delta \cdot \varphi \left(\frac{m}{(V - V_c)^{\frac{1}{2}}} \right) \quad (3.60)$$

with universal scaling function

$$\varphi(x) = \frac{1}{3} - 2Q \cdot \frac{1}{x^2} \quad (3.61)$$

and the new critical exponent δ. The mean field value of δ is $\delta = 3$.

Thus, if you plot (h/m^3) against $(m/(V - V_c)^{\frac{1}{2}})$, curves for different values of V overlap when varying m according to $h = \operatorname{artanh}(m) - 2 \cdot QVm$, and all are similar to the scaling function φ. In Fig. 3.13 a) for three values $V = V_c + 0.02$, $V = V_c + 0.06$, $V = V_c + 0.12$, the curves are plotted and compared to the scaling function φ in the form $(h/m^3) = (1/3) - 2Q((m/(V - V_c)^{\frac{1}{2}})^{-2}) = \varphi(m/(V - V_c)^\beta)$ in dashed lines.

In Fig. 3.13 b) the critical exponent δ is investigated in the double logarithmic plot

$$\ln(h) = \delta \cdot \ln(m) + \ln(k) \quad (3.62)$$

with a fixed value of k as

$$k := \frac{m}{(V(m) - V_c)^\beta} = 10.0 \quad (3.63)$$

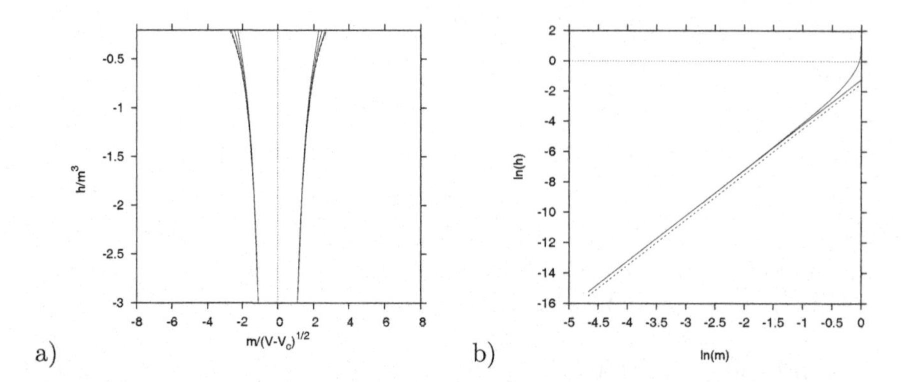

Fig. 3.13 The universal scaling function a) is asymptotically reached, here for several curves with different values of the coupling strength V. b) The critical exponent is also asymptotically reached, again for various values of V and compared with a straight line (dashed) of slope $\delta = 3$.

hence for varying m the value of V is given by

$$V = \left(\frac{k}{m}\right)^{-\frac{1}{\beta}} + V_c \tag{3.64}$$

giving the slope $\delta = 3$, if using $\beta = 1/2$. In Fig. 3.13 b) the dashed line gives the slope $\delta = 3$ (here as $3 \cdot \ln(m) - 1.5$). The full straight line next to the dashed is obtained from $h = \text{artanh}(m) - 2 \cdot QVm$. The full line which is curved upwards is derived from $h = m + \frac{1}{3}m^3 - 2 \cdot QVm$.

3.4.4 *The magnetic susceptibility or second moment of the magnetization diverges*

The second moment of the distribution of spins is historically called magnetic susceptibility χ and obtained via the partition function Z as follows

$$\chi := \langle \sum_{i=1}^{N} \sum_{j=1}^{N} \sigma_i \sigma_j \rangle - \langle \sum_{i=1}^{N} \sigma_i \rangle^2 = \frac{\partial^2 \ln Z}{\partial h^2} \tag{3.65}$$

and then given by the Gibbs free energy in the following way

$$\chi = \left. \frac{N^2}{\frac{\partial^2 \Gamma}{\partial m^2}} \right|_{m_{equilibrium}} \tag{3.66}$$

with

$$\frac{\partial^2 \Gamma}{\partial m^2} = N \cdot \frac{\partial h}{\partial m} \quad . \tag{3.67}$$

Here we have in mean field

$$\frac{\partial^2 \Gamma}{\partial m^2} = -2QVN + N\mathrm{artanh}'(m_e) \tag{3.68}$$

where $m_{equilibrium}$ or simply m_e is again given by Newton's method from the self-consistency equation. Furthermore, $\mathrm{artanh}'(m_e) = 1/(1 - m_e^2)$.

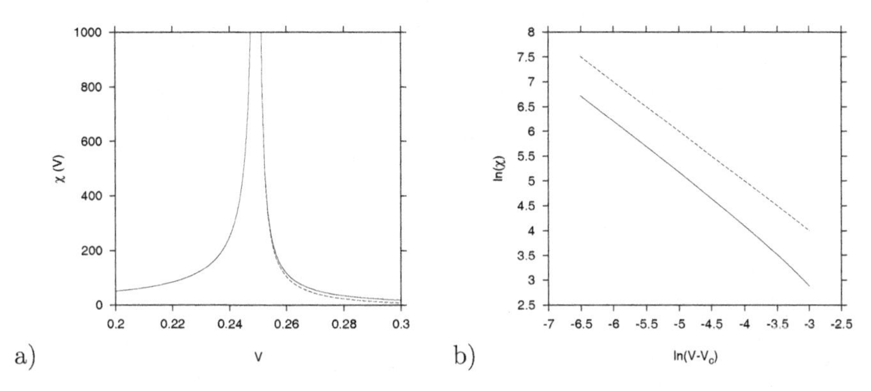

Fig. 3.14 a) The variance of the total magnetization diverges at the critical coupling strength V_c with power laws on both sides. The full curve gives the solution to Eq. (3.66), whereas the dashed line shows the power law shape. b) For $V > V_c$ the scaling plot shows good convergence to the straight line of $\gamma = -1$ for small differences $V - V_c$.

For $V < V_c$ we obtain

$$\chi = \frac{N}{\left(1 - \frac{V}{V_c}\right)} \tag{3.69}$$

since $m_e = 0$. This curve diverges with exponent -1, defining the critical exponent $\gamma = 1$, when V increases to V_c.

For $V > V_c$ we obtain

$$\chi = \frac{N}{2\left(\frac{V}{V_c} - 1\right)} \tag{3.70}$$

with m_e from the zeros of $h = \mathrm{artanh}(m) - 2QVm$. This curve diverges with exponent -1, hence critical exponent $\gamma = 1$, when V decreases from above to V_c. The dashed curve in Fig. 3.14 a) on the right hand side of

V_c comes from Taylor's expansion, hence $h \approx m + (1/3)m^3 - 2QVm$. The calculation is as follows

$$\begin{aligned}\chi &= N\left(\left(1 - \frac{V}{V_c}\right) + m_e^2\right)^{-\gamma} \\ &\approx N\left(\left(1 - \frac{V}{V_c}\right) + \left(3\left(\frac{V}{V_c} - 1\right) - \frac{3}{4}m_e^4 + ...\right)\right)^{-\gamma}\end{aligned} \tag{3.71}$$

with m_e^4 and higher terms going to zero for $V \to V_c$.

To obtain the critical exponent γ directly, the double logarithmic plot has to be considered as shown in Fig. 3.14b) here for $V > V_c$. The upper curve gives a line of slope -1 (here $-1 \cdot \ln(V - V_c) + 1.0$), the full line gives $\ln(\chi)$ as function of $\ln(V - V_c)$ with m_e from the zeros of equation Eq. (3.46). The lower dashed line uses Taylor's expansion $h = m + (1/3)M^3 - 2 \cdot QVm$ for m_e. The plots are in the form

$$\ln\chi = \ln\left(\frac{N}{\left(-2QV + \frac{1}{1-m_e^2}\right)}\right) \tag{3.72}$$

with m_e in the forms just discussed. The critical exponent γ for $V > V_c$ is given by the following expression

$$\chi \sim \left(\frac{V}{V_c} - 1\right)^{-\gamma} \tag{3.73}$$

or $\tilde{\gamma}$ with $V < V_c$

$$\chi \sim \left(1 - \frac{V}{V_c}\right)^{-\tilde{\gamma}} \tag{3.74}$$

on the other side of the susceptibility curve. The exponents γ and $\tilde{\gamma}$ are equal and so do not have to be distinguished here.

3.4.5 *State equation in mean field*

The equation of state for any general theory defined by, for example, a partition function, and from that given Helmholtz free energy and Gibbs free energy $\Gamma(m, V)$, is given by

$$h := \left(\frac{\partial \Gamma}{\partial m}\right)_V \tag{3.75}$$

and we find ourselves back at the self-consistency equation in the case of mean field

$$h = \text{artanh}(m) - 2 \cdot QVm \tag{3.76}$$

and its approximation in Taylor's expansion of artanh(m)

$$h = m + \frac{1}{3}m^3 - 2 \cdot QVm \quad . \tag{3.77}$$

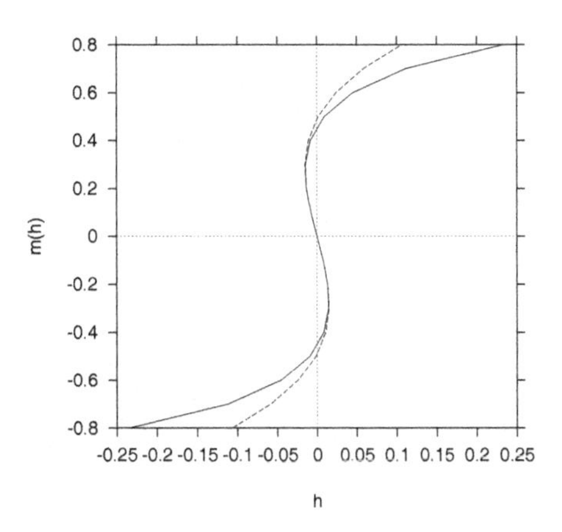

Fig. 3.15 Solution of equation of state, which is again the mean field self-consistency equation (full line) and the approximation to third order in m.

Figure 3.15 compares the actual curves for $h = \text{artanh}(m) - 2 \cdot QVm$ (full line) and its approximation $h = m + \frac{1}{3}m^3 - 2 \cdot QVm$ (dashed line).

Surprisingly, in many ways the $p(M)$ behavior and the analytic behavior of the mean field $-\ln(\Gamma)$ Gibbs free energy are more similar than the thermodynamic Gibbs free energy, in terms of two maxima separated by a valley. For further discussions on this see [Zinn-Justin (1989)].

Finally, we use the mean field state equation to plot $m(V, H)$ as a function of the natural independent variables of the Ising model, namely V and h (see Fig. 3.16). For each value of V and h we search for the zeros of $f(m) := \text{artanh}(m) - 2 \cdot QVm - h$ via Newton's method. Since we are interested in the solution with highest magnetization, we choose as the starting value for the Newton iteration $m_{n+1} = m_n - f(m_n)/f'(m_n)$ a high value of m, e.g. $m_0 = 0.8$. The derivative of f is $f'(m) = -2QV + 1/(1+m)$.

Due to this method of searching for a specific solution to the state equation we have implicitly achieved to select the zero external magnetic

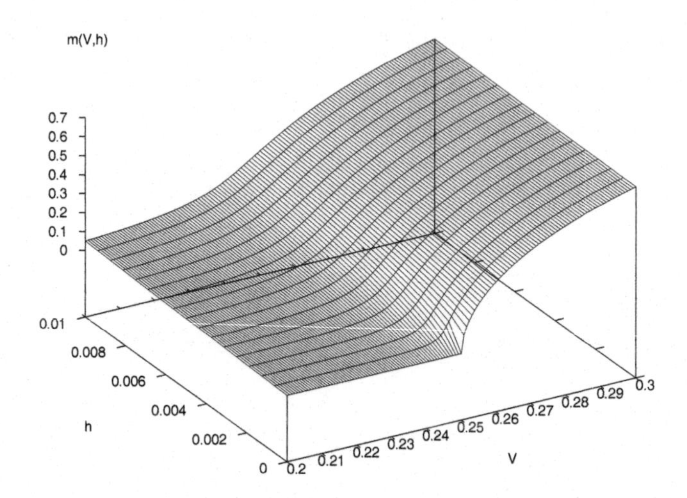

Fig. 3.16 The state equation in its natural variables $m(V, H)$ as function of V and h as the independent variables. The magnetization m is obtained via Newton's method from $\text{artanh}(m) - 2 \cdot QVm - h =: f(m) = 0$. The critical threshold for vanishing external magnetic field is clearly visible, while the positive external field results in a smoothing of the curves around the threshold.

field solution, $h = 0$, in the limit for positive external field going to zero. The threshold behavior for $h \to 0$ of $m(V)$ is clearly visible in Fig. 3.16. We will come back to this picture in Chapter 4, as there appears in an epidemic model, hence a non-equilibrium system, a similar scenario.

3.5 Critical Exponents of the Ising Model beyond Mean Field

In the last section we determined the basic critical exponents in mean field approximation. Tragically for Ising, the one-dimensional model which he investigated in 1925 does not show both sides of the phase transition, i.e. the critical coupling strength is infinite [Ising (1925)]. So in that analytically solvable model in one dimension (see [Yeomans (1992)] for a recent demonstration of the proof) only the one-humped phase is visible, whereas most critical exponents are only defined on the other side of the phase transition.

In 1944 an analytic solution for the two-dimensional Ising model could finally be given for the case where no external magnetic field is present

[Onsager (1944)]. As of today neither the two-dimensional Ising model with external field nor the three-dimensional Ising model, which corresponds to physical systems under investigation, can be solved.

In three dimensions there are only numerical estimates for the critical exponents obtained under heavy computional effort [Landau and Binder (2000)]. Since the complete enumeration is not possible for large enough system sizes we only have stochastic simulations; namely the Metropolis algorithm which is similar to the Glauber model but with different transition rates giving the same stationary state. Around criticality the broad maximum needs long simulation times to be explored well, and the difference of behavior just below and just above criticality is difficult to discern even after long simulations.

For the two-dimensional Ising model, the analytic solution [Onsager (1944)] reveals three critical exponents; namely $\alpha = 0$, $\beta = 1/8$ and $\gamma = 7/4$ (see e.g. [Landau and Binder (2000)], page 17). From general arguments for critical phenomena, not all critical exponents are independently defined but there are relations between them, that are nowadays accepted as equations; namely the Rushbrooke equation

$$\alpha + 2\beta + \gamma = 2 \tag{3.78}$$

in agreement with the above given three exponents, and the hyperscaling relation (see e.g. [Landau and Binder (2000)], page 19)

$$d \cdot \nu = 2 - \alpha \tag{3.79}$$

with dimensionality d of the system. It follows that $\nu = 1$. Furthermore, $\gamma = (2 - \eta)\nu$, $\gamma = \beta(\delta - 1)$ (see [Yeomans (1992)], page 30) gives $\delta = 15$ and $\eta = 1/4$ for the two-dimensional Ising model.

Besides the static critical exponents, which are obtained from the stationary state of the Ising model, the dynamic version, e.g. the Glauber dynamics, also shows dynamic exponents. In particular, it shows the exponent z defined by the magnetization of the Ising model decaying to zero at the critical coupling strength V_c with a power law

$$m(t) \sim t^{-\frac{\beta}{z\nu}} \tag{3.80}$$

from any non-zero magnetization state (see [Landau and Binder (2000)], page 100). Here β and ν are exactly known for the two-dimensional Ising model as seen above, with the spatial correlation length $\xi \sim |V - V_c|^{-\nu}$. A very good and long standing numeric value was given by [Stauffer (1997)] with $z = 2.18$, whereas it had been argued before that $z = 2$ would be the correct value. Very recently research effort has gone into the question if $z = 2$ or significantly differ (see [Tiggemann (2004)] with $z = 2.167 \neq 2$). Hence, it is not surprising that in the case of spatial birth–death processes, where in one dimension already a non-trivial phase transition is present, but still no analytic solution available, the determination of critical exponents is computationally difficult and up for debate even today.

Chapter 4

Partial Immunization Models

In the context of multi-strain models of pathogens infecting a host population, models have recently arisen for partial immunization. A host being infected with one strain is assumed to be only partially immune to another strain, proportional to the genetic distance between the strains. Looking at simple models of reinfection with partial immunization, a so-called reinfection threshold has been postulated and subsequently debated.

We saw in Chapter 2 that the birth–death process or SIS epidemic has a critical transition, directed percolation. The SIR epidemic, sometimes also called a general epidemic, also has a transition with characteristics of ordinary bond percolation [Grassberger (1983)]. Models with partial immunization somehow interpolate between these two basic epidemiological models, and are of interest to theoretical investigation for that reason.

It turns out that the models with partial immunization that have recently been investigated by statistical physicists are identical to those where the reinfection threshold has been observed. It turns out that the reinfection threshold is the mean field fingerprint of a transition, which in spatial, e.g. two-dimensional, stochastic systems is a transition between annular growth, typical for the SIR system above its criticality threshold, and compact growth, typical for the SIS system above its criticality threshold.

We will now investigate a spatial master equation for the simplest epidemiological system with reinfection, the SIRI system. We will characterize it as of qualitatively identical to the statistical physics models of partial immunization and, by showing the same universal behavior between the two, derive the mean field equations for the mean total numbers of hosts in the various classes. The mean field equations show the reinfection threshold as a clear threshold and locate it at the transition between annular growth and compact growth in the underlying spatial system.

4.1 A Model with Partial Immunization: SIRI

In the the SIRI-model we have as transition rates the infection β as a transition from S to I, the recovery rate γ from I into the R class, and as a special ingredient for the only partial immunization or reinfection $\tilde{\beta}$. The reaction schemes for transitions of the SIRI system is given by

$$\begin{aligned} S + I &\xrightarrow{\beta} I + I \\ I &\xrightarrow{\gamma} R \\ R + I &\xrightarrow{\tilde{\beta}} I + I \end{aligned}$$

with S susceptible, I infected, R recovered. For the SIRI model we have in the spatial set up for the variables S_i, I_i and $R_i \in \{0, 1\}$ the constraint that an individual i belongs to one of the three classes

$$S_i + I_i + R_i = 1 \quad . \tag{4.1}$$

Analogously to the SIS system in Chapter 2 the master equation for the probability $p(S_1, I_1, R_1, S_2, I_2, R_2, ..., R_N, t)$ of the spatial SIRI system is for N individuals

$$\begin{aligned} &\frac{d}{dt}\, p\ (S_1, I_1, R_1, S_2, I_2, R_2, ..., R_N, t) \\ &= \sum_{i=1}^{N} \beta \left(\sum_{j=1}^{N} J_{ij} I_j \right) (1 - S_i)\ p(S_1, I_1, R_1, ..., 1 - S_i, 1 - I_i, R_i ..., R_N, t) \\ &+ \sum_{i=1}^{N} \gamma\ (1 - I_i)\ p(S_1, I_1, R_1, ..., S_i, 1 - I_i, 1 - R_i ..., R_N, t) \\ &+ \sum_{i=1}^{N} \tilde{\beta} \left(\sum_{j=1}^{N} J_{ij} I_j \right) (1 - R_i)\ p(S_1, I_1, R_1, ..., S_i, 1 - I_i, 1 - R_i ..., R_N, t) \\ &- \sum_{i=1}^{N} \left[\beta \left(\sum_{j=1}^{N} J_{ij} I_j \right) S_i + \gamma\ I_i + \tilde{\beta} \left(\sum_{j=1}^{N} J_{ij} I_j \right) R_i \right]\ p(...S_i, I_i, R_i ...) \end{aligned} \tag{4.2}$$

and formulated in terms of the three variables S_i, I_i and $R_i \in \{0, 1\}$. The adjacency matrix (J_{ij}) contains 0 for no connection and 1 for a connection between individuals i and j; hence $J_{ij} = J_{ji} \in \{0, 1\}$ for $i \neq j$ and $J_{ii} = 0$.

A first simulation of the model can be seen in Fig. 4.1. For the simulation of the master equation the so-called Gillespie algorithm was used, giving exponential waiting times between transitions [Gillespie (1976)] and [Gillespie (1978)]. Hence time is here a real variable.

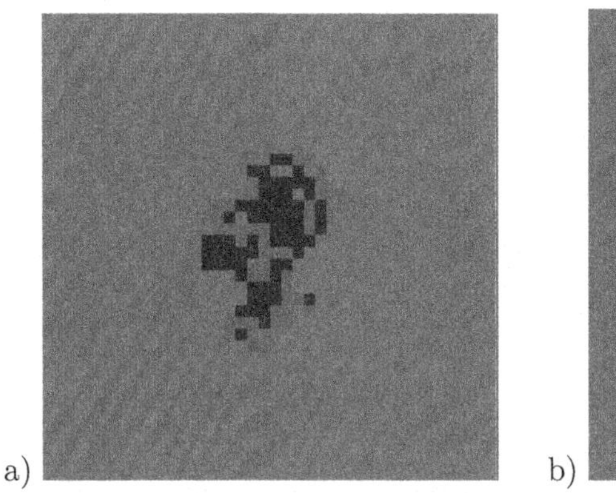

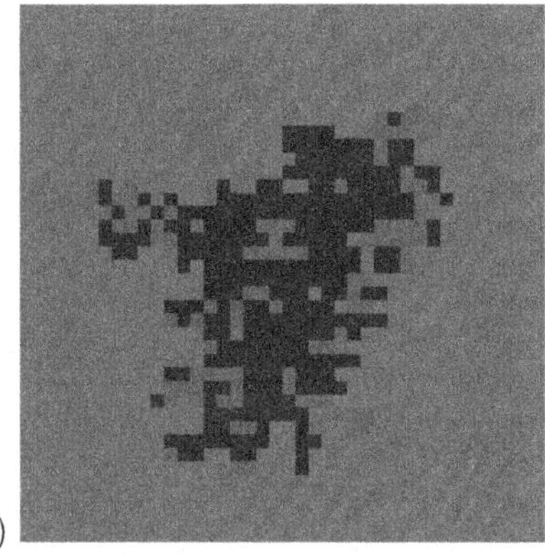

Fig. 4.1 Simlation of a two-dimensional SIRI spreading experiment. Light S, and initially one infected I (medium) at the center of the lattice spreads the infection, leaving mostly recovered R (dark) behind. Parameters of the simulation are $\beta = 1$, $\gamma = 1$ and $\tilde{\beta} = 0.3$, hence also the ratio $\sigma := \tilde{\beta}/\beta = 0.3$.

To obtain expectation values, i.e. for the total number of infected hosts at a given time

$$\langle I \rangle(t) := \sum_{S_1=0}^{1} \sum_{I_1=0}^{1} \sum_{R_1=0}^{1} \sum_{S_2=0}^{1} \cdots \sum_{R_N=0}^{1} \left(\sum_{i=1}^{N} I_i \right) p(S_1, I_1, R_1, S_2, ..., R_N, t) \tag{4.3}$$

we can take the constraint $S_i + I_i + R_i = 1$ Eq. (4.1) into account by replacing $S_i = 1 - I_i - R_i$ by the two independent variables I_i and R_i, giving

$$\langle I \rangle(t) := \sum_{I_1=0}^{1} \sum_{R_1=0}^{1-I_i} \sum_{I_2=0}^{1} \cdots \sum_{R_N=0}^{1-I_N} \left(\sum_{i=1}^{N} I_i \right) p(I_1, R_1, I_2, ..., R_N, t) \quad . \tag{4.4}$$

We also take into account that by fixing I_i to either zero or one, we only need to consider for R_i those possibilities that are left, either zero or one, if $I_i = 0$, or just zero if $I_i = 1$. This Eq. (4.4) might be the easier form to use for the following analytic calculations, while Eq. (4.3) might give results which are easier to interpret. Both forms give, of course, equivalent results.

Thus the master equation becomes in independent variables I_i and R_i

$$
\begin{aligned}
\frac{d}{dt}\, p\ (I_1, R_1, I_2, R_2, ..., R_N, t) \\
&= \sum_{i=1}^{N} \beta \left(\sum_{j=1}^{N} J_{ij} I_j \right) (I_i + R_i)\ p(I_1, R_1, ..., 1 - I_i, R_i ..., R_N, t) \\
&+ \sum_{i=1}^{N} \gamma\, (1 - I_i)\ p(I_1, R_1, ..., 1 - I_i, 1 - R_i ..., R_N, t) \\
&+ \sum_{i=1}^{N} \tilde{\beta} \left(\sum_{j=1}^{N} J_{ij} I_j \right) (1 - R_i)\ p(I_1, R_1, ..., 1 - I_i, 1 - R_i ..., R_N, t) \\
&- \sum_{i=1}^{N} \left[\beta \left(\sum_{j=1}^{N} J_{ij} I_j \right) (1 - I_i - R_i) + \gamma\, I_i + \tilde{\beta} \left(\sum_{j=1}^{N} J_{ij} I_j \right) R_i \right] \\
&\qquad \cdot p(I_1, ..., I_i, R_i, ... R_N, t) \quad .
\end{aligned}
\tag{4.5}
$$

This form allows for straightforward evaluation as in Chapter 2 for the SIS case.

To calculate the dynamics of the moments, mean total number of infected etc., we take the definition of the moments (e.g. Eq. (4.4)) and its time derivative

$$
\frac{d}{dt} \langle I \rangle (t) = \sum_{I_1=0}^{1} \sum_{R_1=0}^{1-I_i} \sum_{I_2=0}^{1} \cdots \sum_{R_N=0}^{1-I_N} \left(\sum_{j=i}^{N} I_i \right) \frac{d}{dt} p(I_1, R_1, I_2, ..., R_N, t) \quad .
\tag{4.6}
$$

Now we must insert the master equation into Eq. (4.6) for the expression $\frac{d}{dt} p(I_1, R_1, I_2, ..., R_N, t)$ and after some calculation, parallel to the extended calculations in Chapter 2 for the SIS epidemic (although with more writing effort due to the larger number of variables of the system) we find in terms of all variables S, I and R

$$\frac{d}{dt}\langle S\rangle = -\beta\,\langle SI\rangle_1$$

$$\frac{d}{dt}\langle I\rangle = \beta\,\langle SI\rangle_1 - \gamma\langle I\rangle + \tilde{\beta}\,\langle RI\rangle_1 \tag{4.7}$$

$$\frac{d}{dt}\langle R\rangle = \gamma\langle I\rangle - \tilde{\beta}\,\langle RI\rangle_1$$

where e.g.

$$\langle RI\rangle_1(t) := \sum_{S_1=0}^{1}\sum_{I_1=0}^{1}\ldots\sum_{R_N=0}^{1}\left(\sum_{i=1}^{N}\sum_{j=1}^{N}(J^1)_{ij}R_iI_j\right)p(S_1,I_1,R_1,...,R_N,t) \tag{4.8}$$

is the mean number of pairs of recovered next to infected. In the equation for the dynamics of $\langle RI\rangle_1$ an expression $\langle RI\rangle_2$ could also show up, meaning power of two of the adjacency matrix J^2 and then its elements $(J^2)_{ij}$.

In mean field approximation (see Chapter 2 for details) we obtain

$$\langle RI\rangle_1(t) \approx \frac{Q}{N}\langle R\rangle\cdot\langle I\rangle \tag{4.9}$$

etc. such that the equation system Eq. (4.7) gives a closed ODE system of the expected form. However, in some areas of the parameter space, especially close to phase transitions, this mean field approximation becomes increasingly less accurate, until at the phase transition it is often very misleading (in the sense that in the Ising model in one dimension, in mean field a phase transition is predicted, where in the analytical solution it turned out not to have one [Ising (1925)].

The mean field equations for the SIRI model, therefore, are

$$\frac{d}{dt}\langle S\rangle = -\beta\frac{Q}{N}\,\langle S\rangle\langle I\rangle$$

$$\frac{d}{dt}\langle I\rangle = \beta\frac{Q}{N}\,\langle S\rangle\langle I\rangle - \gamma\langle I\rangle + \tilde{\beta}\frac{Q}{N}\,\langle R\rangle\langle I\rangle \tag{4.10}$$

$$\frac{d}{dt}\langle R\rangle = \gamma\langle I\rangle - \tilde{\beta}\frac{Q}{N}\,\langle R\rangle\langle I\rangle \quad .$$

In the ODE system Eq. (4.10) one can study stationary solutions when including birth into the susceptible class and death from all classes. In the stochastic system then even, the formal stationary solution only gives disease-free solutions as an absorbing state, since there is always a small probability of extinction in the I class. Hence spreading experiments or the steady import of infected from outside (and then considering the system in the limit of vanishing external import) can help to some extent [Lübeck and Willmann (2002)]). See also [Dickman and Vidigal (2002)] for another approach facing the problem of an absorbing boundary to obtain a quasi-stationary solution.

In the general system without the approximation of mean field behavior, Eq. (4.7), one has to find a dynamic for the pairs $\langle SI\rangle_1$ and $\langle RI\rangle_1$. This will include triples $\langle SIS\rangle_{11}$ etc. for which dynamics have to be found in turn, then including even higher moments (see Chapter 2 for details of the calculations). Pair approximation is the next step after mean field to close the system. In the pair approximation, triples are approximated by an expression which only has means and pairs.

Taking again the constraint $S_i + I_i + R_i = 1$ into account and replacing the set of variables S, I, R with the independent variables I and R hence $S_i = 1 - I_i - R_i$ we obtain for the pairs' expectation value $\langle SI\rangle_1$

$$\langle SI\rangle_1 = \langle\sum_{i=1}^{N}\sum_{j=1}^{N} J_{ij} S_i I_j\rangle$$

$$= \langle\sum_{i=1}^{N}\sum_{j=1}^{N} J_{ij}(1 - I_i - R_i) I_j\rangle$$

$$= \langle\sum_{j=1}^{N} I_j \sum_{j=1}^{N} J_{ij}\rangle - \langle II\rangle_1 - \langle RI\rangle_1$$

and with

$$\sum_{j=1}^{N} J_{ij} =: Q_i \tag{4.11}$$

we obtain the number of neighbors of site i being equal for all i in regular lattices. Hence $Q_i = Q$, where Q is often also called the coordination number, we obtain

$$\langle SI\rangle_1 = Q\langle I\rangle - \langle II\rangle_1 - \langle RI\rangle_1 \quad . \tag{4.12}$$

Inserting this Eq. (4.12) into the dynamics for the mean values Eq. (4.7) we obtain

$$\frac{d}{dt}\langle I\rangle = (\beta Q - \gamma)\langle I\rangle - \beta\,\langle II\rangle_1 + (\tilde{\beta} - \beta)\,\langle RI\rangle_1$$
$$\frac{d}{dt}\langle R\rangle = \gamma\langle I\rangle - \tilde{\beta}\,\langle RI\rangle_1 \tag{4.13}$$

and as a first result we recover for $\tilde{\beta} = \beta$, i.e. where the reinfection is equal to the initial infection, the solution obtained in the SIS system

$$\frac{d}{dt}\langle I\rangle = (\beta Q - \gamma)\langle I\rangle - \beta\,\langle II\rangle_1 \quad . \tag{4.14}$$

The other limit of the SIRI model, the SIR limit $\tilde{\beta} = 0$, can, of course, be obtained from Eq. (4.7) directly by inserting $\tilde{\beta} = 0$.

4.2 Local Quantities

Finally, considering local quantities like $\langle I_i\rangle(t)$ which, in a countinuous space model, would correspond to the local density $\rho(x,t)$ with spatial variable x corresponding to i and lattice spacing a from our lattice model going to zero, we obtain

$$\frac{d}{dt}\langle I_i\rangle = \beta\sum_{j=1}^{N} J_{ij}\langle S_i I_j\rangle - \gamma\langle I_i\rangle + \tilde{\beta}\sum_{j=1}^{N} J_{ij}\langle R_i I_j\rangle \quad . \tag{4.15}$$

Alternatively, in independent variables I_i and R_i we obtain

$$\frac{d}{dt}\langle I_i\rangle = \beta\nabla^2\langle I_i\rangle + (\beta Q - \gamma)\langle I_i\rangle - \beta\sum_{j=1}^{N} J_{ij}\langle I_i I_j\rangle + (\tilde{\beta} - \beta)\sum_{j=1}^{N} J_{ij}\langle R_i I_j\rangle \tag{4.16}$$

where we use the discrete version of the diffusion operator

$$\nabla^2\langle I_i\rangle = \sum_{j=1}^{N} J_{ij}(\langle I_j\rangle - \langle I_i\rangle) \tag{4.17}$$

The discretized diffusion operator [Cardy and Täuber (1998)] for any local function f_i is defined as

$$\nabla^2 f_i := \sum_{j=1}^{N} J_{ij}(f_j - f_i) \tag{4.18}$$

which in one dimension simply gives

$$\nabla^2 f_i = J_{i,i-1}(f_{i-1} - f_i) + J_{i,i+1}(f_{i+1} - f_i) = f_{i+1} - 2 \cdot f_i + f_{i-1} \quad . \tag{4.19}$$

Thus, the dynamics for both mean values is given by

$$\begin{aligned} \frac{d}{dt}\langle I_i \rangle &= \beta \nabla^2 \langle I_i \rangle + (\beta Q - \gamma)\langle I_i \rangle - \beta \sum_{j=1}^{N} J_{ij}\langle I_i I_j \rangle + (\tilde{\beta} - \beta) \sum_{j=1}^{N} J_{ij}\langle R_i I_j \rangle \\ \frac{d}{dt}\langle R_i \rangle &= \gamma \langle I_i \rangle - \tilde{\beta}\, \langle R_i I_i \rangle_1 \quad . \end{aligned} \tag{4.20}$$

Further equations for $\frac{d}{dt}\langle I_i I_j \rangle$ and $\frac{d}{dt}\langle R_i I_j \rangle$ must be calculated explicitly from the master equation.

In Eq. (4.20) the drift and diffusion terms, well known from PDEs, are recovered along with the expected reaction terms. Of course, creation from one particle at one site to a neighboring site and subsequent annihilation on the original site can be interpreted as diffusion (see remark in [Grassberger and de la Torre (1979)]).

In the spatial system we have in two and higher dimensions annular growth near the SIR-limit of the SIRI system and compact growth for supercritical values around the SIS limit [Grassberger, Chaté and Rousseau (1997)]. There is also a field theoretic equation system given, which can be compared with our Eq. (4.20).

Whereas [Grassberger, Chaté and Rousseau (1997)]) consider a two-dimensional spatial system (and briefly mention the degenerate one-dimensional case), [Dammer and Hinrichsen (2004)] also investigate the partial immunization system for three to six spatial dimensions, including the phase diagrams. For the SIS epidemic Eq. (4.20) would give the widely investigated Kolmogorov–Fisher equation [Brockmann and Hufnagel (2007); Boto and Stollenwerk (2009)] where traveling waves can be observed, even in superdiffusive contact networks.

4.3 Dynamics Equations for Global Pairs

A more detailed calculation gives the dynamics of the pairs $\langle RI \rangle_1$ and $\langle II \rangle_1$ as

$$\frac{d}{dt}\langle RI \rangle_1 = \gamma\langle II \rangle_1 + \beta\langle RI \rangle_2 - \gamma\langle RI \rangle_1 + (\tilde{\beta}-\beta)\langle RRI \rangle_{1,1} - \beta\langle RII \rangle_{1,1} - \tilde{\beta}\langle IRI \rangle_{1,1} \tag{4.21}$$

where we have now higher-order terms involved like

$$\langle RI \rangle_1(t) := \sum_{S_1=0}^{1} \dots \sum_{R_N=0}^{1} \left(\sum_{i=1}^{N} \sum_{j=1}^{N} \sum_{k=1}^{N} J_{ij} J_{jk} R_i I_k \right) p(S_1, I_1, R_1, ..., R_N, t) \tag{4.22}$$

and

$$\langle IRI \rangle_{1,1}(t) := \sum_{S_1=0}^{1} \dots \sum_{R_N=0}^{1} \left(\sum_{i=1}^{N} \sum_{j=1}^{N} \sum_{k=1}^{N} J_{ij} J_{jk} I_i R_j I_k \right) p(S_1, ..., R_N, t) \tag{4.23}$$

as triples with an infected on each side of a recovered.

Similarly, we obtain for the dynamics of the $\langle II \rangle_1$ pairs

$$\frac{d}{dt}\langle II \rangle_1 = 2\beta\langle II \rangle_2 - 2\gamma\langle II \rangle_1 - 2\beta\langle III \rangle_{1,1} + 2(\tilde{\beta} - \beta)\langle IRI \rangle_{1,1} \quad , \tag{4.24}$$

and again as a cross check we obtain the SIS case by $\tilde{\beta} = \beta$ (see previous chapter).

Finally, the dynamics of the $\langle II \rangle_1$ pairs are

$$\frac{d}{dt}\langle RR \rangle_1 = 2\gamma\langle IR \rangle_1 - 2\tilde{\beta}\langle RRI \rangle_{1,1} \quad . \tag{4.25}$$

4.3.1 *The SIRI dynamics under pair approximation*

From the general scheme, e.g. for 3 neighbors, S, I, and R the triples can be approximated by

$$\langle SIR \rangle \approx \frac{\langle SI \rangle \cdot \langle IR \rangle}{\langle I \rangle} \tag{4.26}$$

(see Chapter 2). For the SIRI system we obtain, after some calculation

$$\frac{d}{dt}\langle RR \rangle_1 \approx 2\gamma\langle IR \rangle_1 - 2\tilde{\beta}\frac{\langle RR \rangle_1 \cdot \langle RI \rangle_1}{\langle R \rangle} \quad , \tag{4.27}$$

and

$$\frac{d}{dt}\langle II\rangle_1 \approx 2\beta\frac{(Q\langle I\rangle - \langle II\rangle_1 - \langle RI\rangle_1)^2}{(N - \langle I\rangle - \langle R\rangle)} + 2\tilde{\beta}\frac{\langle IR\rangle_1^2}{\langle R\rangle} - 2\gamma\langle II\rangle_1 \tag{4.28}$$

and finally

$$\frac{d}{dt}\langle RI\rangle_1 \approx \gamma\langle II\rangle_1 + \beta\frac{(Q\langle R\rangle - \langle RI\rangle_1 - \langle RR\rangle_1)\cdot(Q\langle I\rangle - \langle II\rangle_1 - \langle RI\rangle_1)}{(N - \langle I\rangle - \langle R\rangle)}$$
$$+\tilde{\beta}\frac{\langle RR\rangle_1\langle RI\rangle_1}{\langle R\rangle} - \gamma\langle RI\rangle_1 - \tilde{\beta}\frac{\langle IR\rangle_1^2}{\langle R\rangle}\,, \tag{4.29}$$

giving a five-dimensional closed ODE system with including the two mean dynamics.

4.3.2 *Balance equations for means and pairs*

From $S_i + I_i + R_i$ it follows immediately that for the means,

$$\langle S\rangle + \langle I\rangle + \langle R\rangle = N \tag{4.30}$$

holds, and from this that

$$\frac{d}{dt}N = 0 = \frac{d}{dt}\langle S\rangle + \frac{d}{dt}\langle I\rangle + \frac{d}{dt}\langle R\rangle \tag{4.31}$$

also holds. A check for the results of the dynamics

$$\frac{d}{dt}\langle S\rangle = -\beta\,\langle SI\rangle_1$$
$$\frac{d}{dt}\langle I\rangle = \beta\,\langle SI\rangle_1 - \gamma\langle I\rangle + \tilde{\beta}\,\langle RI\rangle_1 \tag{4.32}$$
$$\frac{d}{dt}\langle R\rangle = \gamma\langle I\rangle - \tilde{\beta}\,\langle RI\rangle_1$$

is to insert the three equations into Eq. (4.31) and verify that the sum equals zero. In this case it can be confirmed by eye immediately.

For the pair dynamics in all variables S, I and R, however, the check of the balance is not so obvious. The balance equation is now

$$\langle SS\rangle + \langle II\rangle + \langle RR\rangle + 2\langle SI\rangle + 2\langle SR\rangle + 2\langle IR\rangle = N\cdot Q \tag{4.33}$$

which can be obtained by explicitly expressing all terms including variable S in terms of the independent variables I and R: Hence

$$\langle SR \rangle = Q\langle R \rangle - \langle IR \rangle - \langle RR \rangle \quad ; \tag{4.34}$$

further

$$\langle SI \rangle = Q\langle I \rangle - \langle II \rangle - \langle IR \rangle \tag{4.35}$$

and

$$\langle SS \rangle = NQ - 2Q\langle I \rangle - 2Q\langle R \rangle + \langle II \rangle + \langle RR \rangle + 2\langle IR \rangle \quad . \tag{4.36}$$

The pair balance dynamics is now (with $d(N \cdot Q)/dt = 0$)

$$\frac{d}{dt}(N \cdot Q) = \frac{d}{dt}\langle SS \rangle + \frac{d}{dt}\langle II \rangle + \frac{d}{dt}\langle RR \rangle + 2\frac{d}{dt}\langle SI \rangle + 2\frac{d}{dt}\langle SR \rangle + 2\frac{d}{dt}\langle IR \rangle \tag{4.37}$$

which is exactly fulfiled by the following ODE system for the pair dynamics

$$\begin{aligned}
\frac{d}{dt}\langle SS \rangle_1 &= -2\beta\, \langle SSI \rangle_{1,1} \\
\frac{d}{dt}\langle II \rangle_1 &= 2\beta\, \langle ISI \rangle_{1,1} - 2\gamma\langle II \rangle_{1,1} + 2\tilde{\beta}\, \langle IRI \rangle_{1,1} \\
\frac{d}{dt}\langle RR \rangle_1 &= 2\gamma\langle IR \rangle_1 - 2\tilde{\beta}\, \langle RRI \rangle_{1,1} \\
\frac{d}{dt}\langle SI \rangle_1 &= \beta\, \langle SSI \rangle_{1,1} + \tilde{\beta}\, \langle SRI \rangle_{1,1} - \gamma\langle SI \rangle_1 - \beta\, \langle ISI \rangle_{1,1} \\
\frac{d}{dt}\langle RS \rangle_1 &= \gamma\langle SI \rangle_1 - \beta\, \langle RSI \rangle_{1,1} - \tilde{\beta}\, \langle SRI \rangle_{1,1} \\
\frac{d}{dt}\langle RI \rangle_1 &= \gamma\langle II \rangle_1 + \beta\, \langle RSI \rangle_{1,1} + \tilde{\beta}\, \langle RRI \rangle_{1,1} - \gamma\langle IR \rangle_1 - \tilde{\beta}\, \langle IRI \rangle_{1,1}
\end{aligned} \tag{4.38}$$

giving the complete description for the pair dynamics. Only the triples must be approximated now, as described before. It is

$$\langle SRI \rangle \approx \frac{\langle SR \rangle \cdot \langle RI \rangle}{\langle R \rangle} \tag{4.39}$$

$$\langle RSI \rangle \approx \frac{\langle SR \rangle \cdot \langle SI \rangle}{\langle S \rangle} \tag{4.40}$$

$$\langle RRI \rangle \approx \frac{\langle RR \rangle \cdot \langle RI \rangle}{\langle R \rangle} \tag{4.41}$$

$$\langle IRI \rangle \approx \frac{\langle IR \rangle^2}{\langle R \rangle} \tag{4.42}$$

$$\langle ISI \rangle \approx \frac{\langle SI \rangle^2}{\langle S \rangle} \tag{4.43}$$

$$\langle SSI \rangle \approx \frac{\langle SS \rangle \cdot \langle SI \rangle}{\langle S \rangle} \tag{4.44}$$

and has to be inserted into the above given ODE system to obtain closed pair dynamics.

4.4 Mean Field Model: SIRI with Reintroduced Susceptibles

To obtain any useful insight into the mean field stationary state behavior of the SIRI model, we must investigate the model with the additional reintroduction of susceptibles. This can be done in various ways, such as with death from all classes S, I and R and reintroduction to the S class, i.e. birth into susceptibility [Gomes, White and Medley (2004)] or just by taking the original SIR transition α from recovered to susceptibles.

We consider the following reaction scheme

$$S + I \xrightarrow{\beta} I + I$$

$$I \xrightarrow{\gamma} R$$

$$R + I \xrightarrow{\tilde{\beta}} I + I$$

$$R \xrightarrow{\alpha} S$$

with the additional transition from R to I with rate α.

The reaction scheme results in the master equation of the following form

$$
\begin{aligned}
\frac{d}{dt}\, p\ &(S_1, I_1, R_1, S_2, I_2, R_2, ..., R_N, t) \\
&= \sum_{i=1}^{N} \beta \left(\sum_{j=1}^{N} J_{ij} I_j \right) (1 - S_i)\ p(S_1, I_1, R_1, ..., 1 - S_i, 1 - I_i, R_i ..., R_N, t) \\
&\quad + \sum_{i=1}^{N} \gamma (1 - I_i)\ p(S_1, I_1, R_1, ..., S_i, 1 - I_i, 1 - R_i ..., R_N, t) \\
&\quad + \sum_{i=1}^{N} \tilde{\beta} \left(\sum_{j=1}^{N} J_{ij} I_j \right) (1 - R_i)\ p(S_1, I_1, R_1, ..., S_i, 1 - I_i, 1 - R_i ..., R_N, t) \\
&\quad + \sum_{i=1}^{N} \alpha (1 - R_i)\ p(S_1, I_1, R_1, ..., 1 - S_i, I_i, 1 - R_i ..., R_N, t) \\
&\quad - \sum_{i=1}^{N} \left[\beta \left(\sum_{j=1}^{N} J_{ij} I_j \right) S_i + \gamma I_i + \tilde{\beta} \left(\sum_{j=1}^{N} J_{ij} I_j \right) R_i + \alpha R_i \right] \\
&\qquad \cdot p(...S_i, I_i, R_i ...) \quad .
\end{aligned} \tag{4.45}
$$

For the mean total number of susceptible, infected and recovered hosts we obtain

$$
\frac{d}{dt} \langle S \rangle = \alpha \langle R \rangle - \beta\ \langle SI \rangle_1
$$

$$
\frac{d}{dt} \langle I \rangle = \beta\ \langle SI \rangle_1 - \gamma \langle I \rangle + \tilde{\beta}\ \langle RI \rangle_1 \tag{4.46}
$$

$$
\frac{d}{dt} \langle R \rangle = \gamma \langle I \rangle - \alpha \langle R \rangle - \tilde{\beta}\ \langle RI \rangle_1 \tag{4.47}
$$

again involving pairs of infected and susceptibles or pairs of infected and recovered. These have to be either given with their dynamics or be approximated with the mean field assumption.

The mean field approximation is

$$\begin{aligned}
\frac{d}{dt}\langle S\rangle &= \alpha\langle R\rangle - \beta\frac{Q}{N}\langle S\rangle\langle I\rangle \\
\frac{d}{dt}\langle I\rangle &= \beta\frac{Q}{N}\langle S\rangle\langle I\rangle - \gamma\langle I\rangle + \tilde{\beta}\frac{Q}{N}\langle R\rangle\langle I\rangle \qquad (4.48)\\
\frac{d}{dt}\langle R\rangle &= \gamma\langle I\rangle - \alpha\langle R\rangle - \tilde{\beta}\frac{Q}{N}\langle R\rangle\langle I\rangle
\end{aligned}$$

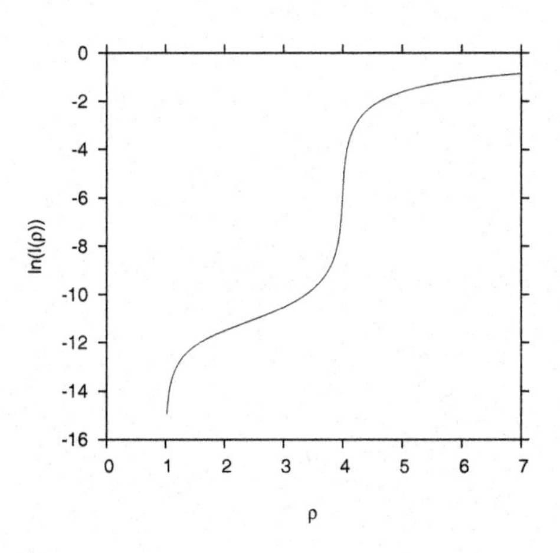

Fig. 4.2 The conventional picture of the reinfection threshold, in semi-logarithmic plot. We use the parameter α instead of the death out of all classes and birth into susceptibles. The same parameter value for $\sigma = 1/4$ is used as in Fig. 4.3, and $\varepsilon = 0.00001$ to demonstrate a clear threshold behavior around $\rho = 1/\sigma$.

This ODE system can be studied in simplified coordinates and the time changed to $\tau := t/\gamma$ so that

$$\rho := \beta Q/\gamma \quad , \qquad (4.49)$$

$\sigma := \tilde{\beta}/\beta$ and

$$\varepsilon := \alpha/\gamma \quad . \qquad (4.50)$$

Further, we consider the densities of susceptibles, infected, and recovered; hence $s := S/N$, $i := I/N$. Then with $R/N = 1 - s - i$ we obtain the two-dimensional ODE system

$$\frac{d}{d\tau}s = \varepsilon(1-s-i) - \rho si$$
$$\frac{d}{d\tau}i = \rho i(s+\sigma(1-s-i)) - i \tag{4.51}$$

The stationary solution is either $i^* = 0 =: i_1^*$ or

$$i^* = -\frac{r}{2} + \sqrt{\frac{r^2}{4} - q} \quad =: i_2^* \tag{4.52}$$

with

$$r := \frac{1}{\rho\sigma}(1 - \rho\sigma + \varepsilon) \tag{4.53}$$

and

$$q := \frac{\varepsilon}{\rho^2\sigma}(1-\rho) \tag{4.54}$$

At $\rho = 1$ the solutions i_1^* and i_2^* meet each other, i.e. $i_2^* = 0$, coinciding with $q = 0$.

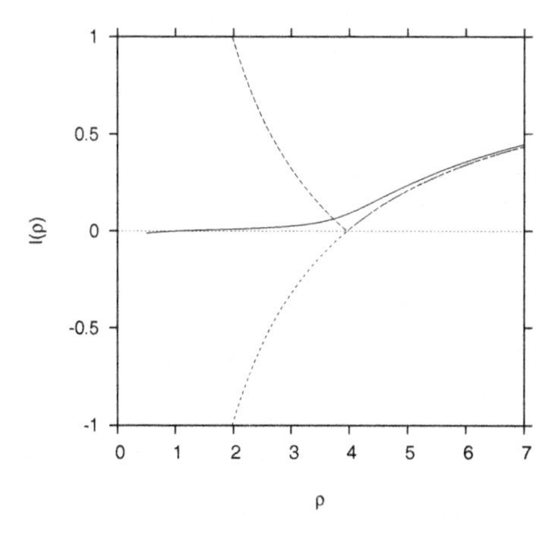

Fig. 4.3 The solution $i_2^* = -\frac{r}{2} + \sqrt{\frac{r^2}{4} - q}$, full line is plotted against the curves $-r$ and its modulus $|r|$. While i_2^* changes from negative to positive at $\rho = 1$, the curves for $-r$ and $|r|$ change at $\rho = 1/\sigma$ for vanishing or small ε. This qualitative change at $\rho = 1/\sigma$ is the reinfection threshold. Parameters are $\sigma = 1/4 = 0.25$ and $\varepsilon = 0.01$.

At $\varepsilon = 0$ we obtain another change of regime with $r = 0$, which is slightly more subtle in the first inspection. The whole concept of this second threshold behavior was questioned in [Breban and Blower (2005)] as a reply to [Gomes, White and Medley (2004)] justified the concept of the reinfection threshold by looking at the behavior of the basic mean field model under vaccination, showing that the first threshold can be shifted towards larger ρ values by the introduction of vaccination, but cannot be shifted beyond the second threshold by any means of vaccination [Gomes, White and Medley (2005)]. Here we demonstrate, that in the SIRI model with reintroduced susceptibles (in our version of the transition from recovered to susceptibles with rate α), the reinfection threshold appears in the limit of α decreasing to zero as a sharp threshold.

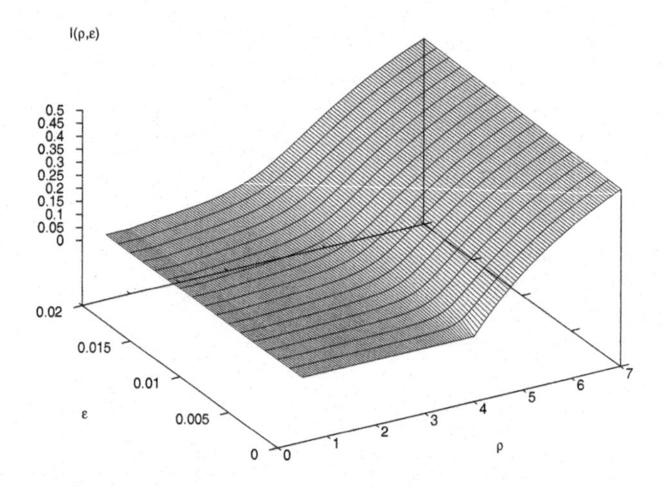

Fig. 4.4 The stationary value of the number of infected individuals with both parameters ρ and ε only shows a threshold behavior at $\rho = 1$ for high ε values, and for vanishing ε the threshold for $\rho = 1/\sigma$. Here in the graphic plot $\rho = 1/\sigma = 4$, where beforehand σ was fixed to be $\sigma := 1/4$.

The threshold behavior is analogous to the Ising spin threshold for example in the limit of vanishing external magnetic field. So we conclude that the reinfection threshold does exist in the sense that any other threshold in physical phase transitions exists or any bifurcation behavior in mean field models exist. From the studies of spatial stochastic epidemics with partial immunization [Grassberger, Chaté and Rousseau (1997); Dammer and Hinrichsen (2004)] we even know that the mean field threshold behavior also qualitatively describes the threshold behavior of spatial models, namely the transition between annular growth and compact growth.

In the following we analyze the mean field behavior of the SIRI model with α-interaction only, and carefully investigate the limiting behavior of vanishing α, finding a sharp transition at $1/\sigma$, the reinfection threshold.

In Fig. 4.3 the solution $i_2^* = -\frac{r}{2} + \sqrt{\frac{r^2}{4} - q}$ (full line) is plotted against the curves $-r$ and its modulus $|r|$. While i_2^* changes from negative to positive at $\rho = 1$, the curves for $-r$ and $|r|$ change at $\rho = 1/\sigma$ for vanishing or small ε. This qualitative change at $\rho = 1/\sigma$ is the reinfection threshold as predicted by [Gomes, White and Medley (2004)].

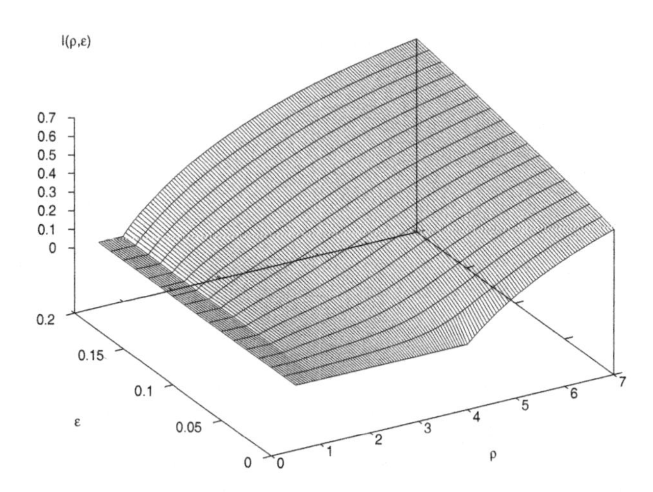

Fig. 4.5 When we look at larger values of ε, here up to $\varepsilon = 0.2$, we also find the first threshold at $\rho = 1$. The continuous change from behavior dominated by the first threshold $\rho = 1$ for $\varepsilon = 0.2$ to the behavior only determined by that surrounding the second threshold $\rho = 1/\sigma$ for $\varepsilon = 0$ can be seen here.

When plotting the stationary state solution as function of both independent parameters ρ and ε (see Fig. 4.4), the threshold behavior for vanishing $\varepsilon = 0$ is clearly visible, whereas for finite ε, the curves for $I^*(\rho)$ are smoothed out around the reinfection threshold.

This behavior is qualitatively very similar to that seen in the magnetic models in Chapter 3. Figure 4.4, which we have just discussed, can be compared to the mean field solution of the Ising model, where the magnetization $m(V, h)$ shows a similar threshold behavior for positive but vanishing external magnetic field $h \to 0$ as shown in Fig. 3.16.

It is interesting to see that for large values of e.g. $\varepsilon = 0.2$ the behavior of I^* is completely dominated by the simple threshold behavior around $\rho = 1$, while the reinfection threshold is not even qualitatively visible (see Fig. 4.5). In contrast, for vanishing $\varepsilon = 0$ there is only the qualitative behavior left from the behavior around the second threshold $\rho = 1/\sigma$. The change between these two extremes is quite continuous, as can be seen in Fig. 4.5. However, close to the reinfection threshold $\rho = 1/\sigma$, the solutions for I^* for small ε are a smoothed out version of that threshold, as better seen in Fig. 4.4.

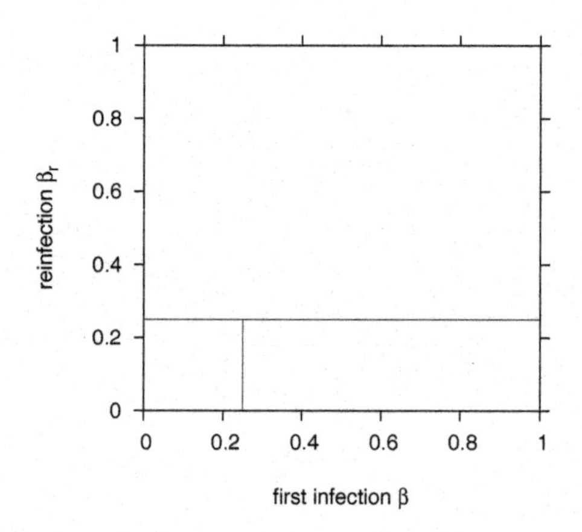

Fig. 4.6 Phase diagram for the mean field model for the two-dimensional case, $Q = 4$ neighbors.

Finally, we look at the phase diagram in the original coordinates, β and $\tilde{\beta}$, as opposed to the changed variables ρ and ε, remembering the definitions for these, Eq. (4.49) and Eq. (4.50). The phase diagram for the mean field model for the two-dimensional case, $Q = 4$ neighbors is shown in Fig. 4.7. The threshold at $\rho = 1$ gives the critical value for β with $\beta = \gamma/Q$, the horizontal line. The threshold for $\rho = 1/\sigma$ gives the critical value for $\tilde{\beta}$ (β_r) with $\tilde{\beta} = \gamma/Q$, the vertical line.

This phase diagram is comparable to the phase diagrams of higher dimensional spatial stochastic simulation where the mean field behavior is approached in about six dimensions [Dammer and Hinrichsen (2004)].

4.4.1 *Pair dynamics for the SIRI model*

We conclude this chapter by mentioning some future extensions of the mean field analysis given above for the SIRI model. The pair dynamics of the model is given by

$$\begin{aligned}
\frac{d}{dt}\langle SS\rangle_1 &= 2\alpha\langle RS\rangle_1 - 2\beta\,\langle SSI\rangle_{1,1} \\
\frac{d}{dt}\langle II\rangle_1 &= 2\beta\,\langle ISI\rangle_{1,1} - 2\gamma\langle II\rangle_1 + 2\tilde{\beta}\,\langle IRI\rangle_{1,1} \\
\frac{d}{dt}\langle RR\rangle_1 &= 2\gamma\langle IR\rangle_1 - 2\tilde{\beta}\,\langle RRI\rangle_{1,1} - 2\alpha\langle RR\rangle_1 \\
\frac{d}{dt}\langle SI\rangle_1 &= \beta\,\langle SSI\rangle_{1,1} + \tilde{\beta}\,\langle SRI\rangle_{1,1} - \gamma\langle SI\rangle_1 - \beta\,\langle ISI\rangle_{1,1} - \alpha\langle RI\rangle_1 \\
\frac{d}{dt}\langle RS\rangle_1 &= \gamma\langle SI\rangle_1 - \beta\,\langle RSI\rangle_{1,1} - \tilde{\beta}\,\langle SRI\rangle_{1,1} + \alpha\langle RR\rangle_1 - \alpha\langle RS\rangle_1 \\
\frac{d}{dt}\langle RI\rangle_1 &= \gamma\langle II\rangle_1 + \beta\,\langle RSI\rangle_{1,1} + \tilde{\beta}\,\langle RRI\rangle_{1,1} \\
&\qquad -\gamma\langle IR\rangle_1 - \tilde{\beta}\,\langle IRI\rangle_{1,1} - \alpha\langle RI\rangle_1
\end{aligned} \tag{4.55}$$

with the usual pair approximation to be used again. Further investigations will show how far the system captures any spatial information, as the SIS pair approximation does (Chapter 2).

4.5 Fruitful Transfer between Equilibrium and Non-Equilibrium Systems

We have seen in the last few chapters, especially in the last sections, how fruitful the transfer of knowledge between notions in equilibrium phase transitions and non-equilibrium phase transitions can be. Since the sixties of the twenties century, dynamic aspects of equilibrium phase transitions have gained increasing attention, while it was not until the seventies that non-equilibrium systems were properly investigated. The directed percolation universality class is the paradigmatic system in this non-equilibrium phase transition research. In Chapter 5 we will take up the parallels between equilibrium and non-equilibrium systems, to explore what can be used to describe the latter further.

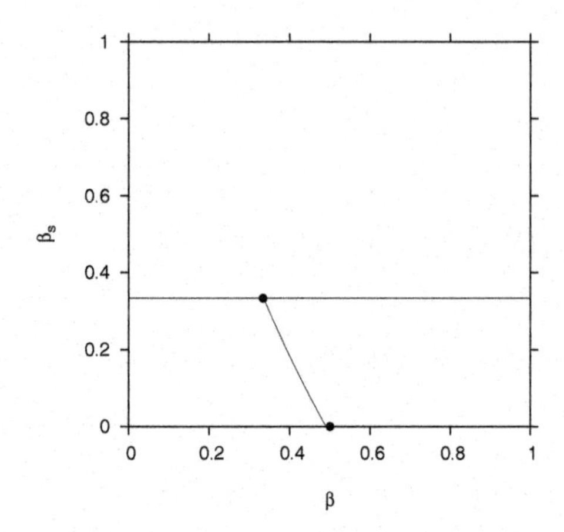

Fig. 4.7 Phase diagram for the pair approximation model for the two-dimensional case, $Q = 4$ neighbors. $\alpha = 0.05$ for the phase transition line between the SIS point and the SIR point.

Chapter 5

Renormalization and Series Expansion: Techniques to Study Criticality

5.1 Introduction

Renormalization gives the final explanation for why, at criticality, only power laws can describe the main features. Power laws are scale free, hence independent of the scale at which you look at the system (see e.g. [Jensen (1998)]).

Scale-free can be explained as follows: A funtion with power law behavior $f(x) := x^a$ behaves similarly on different scales x and $k \cdot x$ in the sense that the quotient

$$\frac{f(kx)}{f(x)} = \frac{(kx)^a}{x^a} = k^a = \text{const.} \tag{5.1}$$

is independent of x. Hence in this sense, power laws lack a characteristic scale, which means that $f(x)$ has the following scaling form

$$f(x) = k^{-a} \cdot f(kx) \quad . \tag{5.2}$$

On the contrary, the exponential function $g(x) = e^{ax}$ has a scale given by the parameter a; hence the quotient

$$\frac{g(kx)}{g(x)} = \frac{e^{a(kx)}}{e^{ax}} = e^{(k-1)ax} \tag{5.3}$$

still depends on x with the scale of ax [Yeomans (1992)].

This self-similarity property at criticality can be used to study the system's behavior in the critical region. For example, the partition function, introduced in Chapter 3 to give all interesting properties of equilibrium thermodynamic properties, is self-similar at criticality.

In the one-dimensional Ising system the partition function even becomes equal at diffierent scales, making it an ideal system for studying the techiques of renormalization [Yeomans (1992)]. For this system it is

$$Z_{N'}(\mathcal{H}') = Z_N(\mathcal{H}) \tag{5.4}$$

for $N/N' = b^d$, where the scaling factor $b = 2$ and dimension $d = 1$. Therefore, the free energy is

$$F(\mathcal{H}') = b^d \cdot F(\mathcal{H}) \tag{5.5}$$

and magnetization

$$M(\mathcal{H}') = \frac{\partial}{\partial h} F(\mathcal{H}') = b^d \frac{\partial}{\partial h} \cdot F(\mathcal{H}) = b^{d+y_2} M(\mathcal{H}) \tag{5.6}$$

or via

$$p^*(\mathcal{H}'(\{\underline{S}\})) = \sum_{\{\underline{\sigma}\}} p(\{\underline{S}\}|\{\underline{\sigma}\}) \cdot p^*(\mathcal{H}(\{\underline{\sigma}\})) \tag{5.7}$$

and then consider $p(M)$.

5.2 Real Space Renormalization in One-Dimensional Lattice Gas

Since we can always transform between spins σ_i and occupation numbers I_i we will now show the renormalization directly for the variable I_i. Hence we speak of a lattice gas, rather than a spin system. The model can also be interpreted as an SIS model, but one for which we have a unique non-trivial stationary state, somehow a bit remote from the classical SIS system, since recovery would now also be dependent not on a single individual but its neighbors. We will come back to the proper SIS model and its treatment in Section 5.3. The model to be considered here has the stationary distribution

$$p^*(I_1, ..., I_N) = \frac{1}{Z}\, e^{\mathcal{H}(I_1,...I_N)} \tag{5.8}$$

with

$$\mathcal{H}(I_1, ...I_N) = J \sum_{i=1}^{N} \sum_{j=1}^{N} J_{ij} I_i I_j + H \sum_{i=1}^{N} I_i + C \cdot N \tag{5.9}$$

which can be expressed by $\mathcal{H}(I_1, ...I_N) = \sum_{i=1}^{N} \mathcal{H}_i(I_i, N_i)$ with neighbors of I_i denoted by $Q_i := \sum_{j=1}^{N} J_{ij} I_j$ by

$$\mathcal{H}_i(I_i, Q_i) = J I_i \sum_{j=1}^{N} J_{ij} I_j + H I_i + C \quad . \tag{5.10}$$

The partition function Z is given by

$$Z := \sum_{I_1=0}^{1} \sum_{I_2=0}^{1} ... \sum_{I_N=0}^{1} e^{\mathcal{H}(I_1,...I_N)} \tag{5.11}$$

and with local $\mathcal{H}_i(I_i, N_i)$

$$Z = \sum_{I_1=0}^{1} \sum_{I_2=0}^{1} ... \sum_{I_N=0}^{1} \prod_{i=1}^{N} e^{J I_i \sum_{j=1}^{N} J_{ij} I_j + H I_i + C} \quad . \tag{5.12}$$

From the partition function, all other thermondynamic variables follow. Thus we can renormalize the partition function and then from the result obtain the information for the behavior of the system in total.

In the one-dimensional case with periodic boundary conditions $I_{N+1} = I_1$ we have

$$Z = \sum_{I_1=0}^{1} \sum_{I_2=0}^{1} ... \sum_{I_N=0}^{1} \prod_{i=1}^{N} e^{J I_i (I_{i-1} + I_{i+1}) + H I_i + C} \tag{5.13}$$

or (easy to see with e.g. $N = 4$ sites)

$$Z = \sum_{I_1=0}^{1} \sum_{I_2=0}^{1} ... \sum_{I_N=0}^{1} \prod_{i=1}^{N} e^{2 J I_i I_{i+1} + \frac{1}{2} H (I_i + I_{i+1}) + C} \quad ; \tag{5.14}$$

hence $\mathcal{H}_i = \mathcal{H}_i(I_i, I_{i+1})$. This is the starting point for the renormalization.

The essential idea is to sum up over all odd sites and leave only the even sites in the partition sum, hence $\sum_{I_1=0}^{1} \sum_{I_2=0}^{1} \cdots \sum_{I_N=0}^{1}$ to $\sum_{I_2=0}^{1} \sum_{I_4=0}^{1} \cdots \sum_{I_N=0}^{1}$ assuming N is even.

The result after some simple calculation is

$$Z = \sum_{I_2=0}^{1} \sum_{I_4=0}^{1} ... \sum_{I_N=0}^{1} \prod_{i=2,4,...}^{N} \left(e^{\frac{1}{2} H (I_i + I_{i+2}) + 2C} + e^{(2J + \frac{1}{2})(I_i + I_{i+2}) + (H + 2C)} \right) \tag{5.15}$$

or after relabeling the sites consecutively $(i, i+2, ... \rightarrow j, j+1, ...)$

$$Z = \sum_{I_1=0}^{1} \sum_{I_2=0}^{1} ... \sum_{I_{\frac{N}{2}}=0}^{1} \prod_{j=1,2,...}^{N/2} \left(e^{\frac{1}{2}H(I_j+I_{j+1})+2C} + e^{(2J+\frac{1}{2}H)(I_j+I_{j+1})+(H+2C)} \right). \tag{5.16}$$

The simple calcualtions are as follows: from Eq. (5.14) we regroup the products to

$$Z = \sum_{I_1=0}^{1} \sum_{I_2=0}^{1} ... \sum_{I_N=0}^{1} \prod_{i=1,3,...}^{N} e^{2JI_i(I_{i-1}+I_{i+1})+HI_i+\frac{1}{2}H(I_{i-1}+I_{i+1})+2C} \tag{5.17}$$

such that $Z_{i-1,i,i+1} := e^{2JI_i(I_{i-1}+I_{i+1})+HI_i+\frac{1}{2}H(I_{i-1}+I_{i+1})+2C}$ as the local partition function takes the odd site and both even sites into account. Then we can sum over the odd sums $\sum_{I_1=0}^{1}$, $\sum_{I_3=0}^{1}$ etc. The central step here is then (easy to see with e.g. $N = 4$ sites) to go from

$$Z = \sum_{I_1=0}^{1} \sum_{I_2=0}^{1} ... \sum_{I_N=0}^{1} \prod_{i=1,3,...}^{N} Z_{i-1,i,i+1} \tag{5.18}$$

to

$$Z = \sum_{I_2=0}^{1} \sum_{I_4=0}^{1} ... \sum_{I_N=0}^{1} \prod_{i=1,3,...}^{N} \left(\sum_{I_i=0}^{1} Z_{i-1,i,i+1} \right) \tag{5.19}$$

taking half the sites, the odd ones, into the new local partition function and only leaving the ensemble sum over even sites remaining. In this way, we have scaled the partition function and with it the whole system from N sites to $N/2$ sites.

The partition function in the form Eq. (5.16) is still the original partition function of the zero'th renormalization step Z_0 with parameter set of the zero'th step $\underline{\mu}_0 := (J_0, H_0, C_0)$. After the first renormalization step is performed, the new partition function Z_1 with new parameter set $\underline{\mu}_1 := (J_1, H_1, C_1)$ again has the original functional form of a one-dimensional system, as given in Eq. (5.14). Hence

$$Z_0 = Z_1 \tag{5.20}$$

with

$$Z_0 = \sum_{I_1=0}^{1} \sum_{I_2=0}^{1} \dots \sum_{I_{\frac{N}{2}}=0}^{1} \prod_{j=1,2,\dots}^{N/2} \left(e^{\frac{1}{2}H_0(I_j+I_{j+1})+2C_0} \right.$$

$$\left. + e^{(2J_0+\frac{1}{2}H_0)(I_j+I_{j+1})+(H_0+2C_0)} \right) \tag{5.21}$$

and

$$Z_1 = \sum_{I_1=0}^{1} \sum_{I_2=0}^{1} \dots \sum_{I_{N/2}=0}^{1} \prod_{j=1,2,\dots}^{N/2} e^{2J_1 I_i I_{i+1} + \frac{1}{2}H_1(I_i+I_{i+1})+C_1} \quad . \tag{5.22}$$

More generally, for renormalization steps n to $n+1$ we require

$$Z_{n+1} = Z_n \tag{5.23}$$

which implies the renormalization mapping of the parameters

$$e^{2J_{n+1}I_i I_{i+1} + \frac{1}{2}H_{n+1}(I_i+I_{i+1})+C_{n+1}} \tag{5.24}$$

$$= \left(e^{\frac{1}{2}H_n(I_j+I_{j+1})+2C_n} + e^{(2J_n+\frac{1}{2}H_n)(I_j+I_{j+1})+(H_n+2C_n)} \right)$$

often called the renormalization flow. This is the mapping

$$\underline{\mu}_{n+1} = \underline{f}(\underline{\mu}_n) \tag{5.25}$$

for the variables J_n, H_n and C_n, once we consider the four cases for which I_i, $I_{i+1} \in \{0, 1\}$.

Since we only have terms $(I_i \cdot I_{i+1})$ and $(I_i + I_{i+1})$, there are three equations left for the three variables J_n, H_n and C_n:

(i) $\underline{I_i = I_{i+1} = 0}$:

$$e^{C_{n+1}} = e^{2C_n} + e^{H_n+2C_n} \tag{5.26}$$

(ii) $\underline{I_i = I_{i+1} = 1}$:

$$e^{2J_{n+1}+H_{n+1}+C_{n+1}} = e^{H_n+2C_n} + e^{4J_n+2H_n+2C_n} \tag{5.27}$$

(iii)
$I_i = 1$, $I_{i+1} = 0$ and $I_i = 0$, $I_{i+1} = 1$ give $(I_i \cdot I_{i+1}) = 0$ and $(I_i + I_{i+1}) = 1$:

$$e^{\frac{1}{2}H_{n+1}+C_{n+1}} = e^{\frac{1}{2}H_n+2C_n} + e^{2J_n+\frac{3}{2}H_n+2C_n} \tag{5.28}$$

Inserting Eq. (5.26) in the form $e^{C_{n+1}} = e^{2C_n}(1+e^{H_n})$ into Eq. (5.28) and squaring gives

$$e^{H_{n+1}} = \frac{e^{H_n}}{(1+e^{H_n})^2}\left(1+2e^{2J_n+H_n}+e^{4J_n+2H_n}\right) \tag{5.29}$$

and inserting into Eq. (5.27) gives

$$e^{2J_{n+1}} = \frac{\left(1+e^{4J_n+H_n}\right)\left(1+e^{H_n}\right)}{1+2e^{2J_n+H_n}+e^{4J_n+2H_n}} \quad . \tag{5.30}$$

With

$$x_n := e^{2J_n} \quad , \quad y_n := e^{H_n} \quad , \quad z_n := e^{C_n} \tag{5.31}$$

as new variables we finally obtain

$$\begin{aligned} x_{n+1} &= \frac{(1+x_n^2 y_n)(1+y_n)}{1+2x_n y_n + x_n^2 y_n^2} \\ y_{n+1} &= \frac{y_n}{(1+y_n)^2} \cdot (1+2x_n y_n + x_n^2 y_n^2) \\ z_{n+1} &= z_n^2(1+y_n) \end{aligned} \tag{5.32}$$

for the renormalization map.

For the Ising spin system in one dimension we only have to replace $\sigma_i = 2I_i - 1$ in the function $\mathcal{H}$ to get that

$$\begin{aligned} \mathcal{H}(\sigma_1, \sigma_2, ..., \sigma_N) &:= K\sum_{i=1}^{N} \sigma_i\sigma_{i+1} + B\sum_{i=1}^{N}\sigma_i + \sum_{i=1}^{N} D \\ &= K\sum_{i=1}^{N}(2I_i - 1)\cdot(2I_{j+1}-1) + B\sum_{i=1}^{N}(2I_i - 1) + \sum_{i=1}^{N} D \end{aligned} \tag{5.33}$$

giving

$$\mathcal{H}(I_1, I_2, ..., I_N) = 4K \sum_{i=1}^{N} I_i I_{i+1} + (2B - 4K) \sum_{i=1}^{N} I_i + \sum_{i=1}^{N} (D + K - B)$$
$$= J \sum_{i=1}^{N} I_i I_{i+1} + H \sum_{i=1}^{N} I_i + \sum_{i=1}^{N} C \tag{5.34}$$

and

$$2J = 4K \quad , \quad H = 2B - 4K \quad , \quad C = D + K - B \ . \tag{5.35}$$

With

$$u_n := e^{-4K_n} \quad , \quad v_n := e^{-2_n} \quad , \quad z_n := e^{-4D_n} \tag{5.36}$$

as new variables we finally obtain for the Ising model

$$u_{n+1} = \frac{u_n(1 + v_n)^2}{(u_n + v_n)(1 + u_n v_n)}$$
$$v_{n+1} = \frac{v_n(u_n + v_n)}{(1 + u_n v_n)} \tag{5.37}$$
$$w_{n+1} = \frac{w_n u_n v_n^2 (1 + v_n)^2}{(u_n + v_n)(1 + u_n v_n)}$$

for the renormalization map [Yeomans (1992)]. Though the one-dimensional Ising model under renormalization is exactly treatable, it does not have strictly two phases, hence is of limited use in studying phase tranitions from both sides. However, the basic ideas can be learned here, since the phase transition is exactly at the boundary of the valid parameter region.

We will now show how to obtain an expression like a partition function for non-equilibrium stochastic systems such as the proper SIS-epidemics.

5.3 Directed Percolation and Path Integrals

For a long time it has been numerically established that simple birth–death processes for mutually excluding particles on a lattice belong in criticality to the universality class of directed percolation [Grassberger and de la Torre (1979)]. However, recent attempts have started to describe such hard-core particles in a field theory [Park, Kim and Park (2000)] and more recently, in a formalism easily treated analytically to obtain such field theories, i.e. bosonic theories [van Wijland (2001)]. Van Wijland uses δ-functions built from bose operators.

We show that the δ-bosons used by [van Wijland (2001)] can mimic the spin 1/2 operators used in [Grassberger and de la Torre (1979)] and derive a path integral which can be compared to those analyzed for directed percolation [Janssen (1981)]. To make the link between such hard-core processes and directed percolation precise is especially important for modeling epidemics, which naturally happen in entities of uninfected or single infected individuals, e.g. in plant epidemics plants on regular lattice points (see e.g. [Stollenwerk and Briggs (2000)]), or in animal and human epidemics on social network lattices (e.g. [Rand (1999)]).

5.3.1 *Master equation of the birth–death process*

One of the simplest and best-studied spatial processes is the birth–death process with birth rate β and death rate α on N sites, each of which can be either inhabited $I := 1$, or empty or solo $S := 1$; hence $I = 0$ (in general $S := I - 1$). In this section, α and β will stand for death and birth rate respectively, since a will be used for annihilators, as is convention in particle and stochastical physics.

Translated into epidemiology, I is the infected, S the susceptible class, β the infection rate, α the recovery. We refer to it as SIS system. The master equation for the spatial SIS system is for N lattice points using the master equation approach for a spatial system in a form as for example used in [Glauber (1963)] for spin dynamics,

$$\frac{d}{dt}p(I_1, ..., I_N, t) = \sum_{i=1}^{N} w_{I_i,1-I_i}(t)\ p(I_1, ..., 1-I_i, ..., I_N, t) - \sum_{i=1}^{N} w_{1-I_i,I_i}(t)\ p(I_1, ..., I_i, ..., I_N, t) \tag{5.38}$$

for $I_i \in \{0, 1\}$ and transition rate

$$w_{I_i,1-I_i} = \beta \left(\sum_{j=1}^{N} J_{ij} I_j \right) \cdot I_i + \alpha \cdot (1 - I_i) \quad , \tag{5.39}$$

and

$$w_{1-I_i,I_i} = \beta \left(\sum_{j=1}^{N} J_{ij} I_j \right) \cdot (1 - I_i) + \alpha \cdot I_i \quad , \tag{5.40}$$

with β birth or infection rate and α death or recovery rate. Here (J_{ij}) is the adjacency matrix containing 0 for no connection and 1 for a connection between sites i and j, hence $J_{ij} = J_{ji} \in \{0, 1\}$ for $i \neq j$ and $J_{ii} = 0$.

The master equation can be transformed into a Schrödinger-like equation using operators common in quantum theory [Grassberger and Scheunert (1980); Peliti (1985)], from which a path integral can be derived for the renormalization analysis.

5.3.2 *Schrödinger-like equation*

The master equation Eq. (5.38) can be written in the form of a linear operator equation

$$\frac{d}{dt} |\Psi(t)\rangle = L |\Psi(t)\rangle \tag{5.41}$$

for a Liouville operator L to be calculated from the master equation Eq. (2.2) and with state vector $|\Psi(t)\rangle$ defined by

$$\begin{aligned} |\Psi(t)\rangle &:= \sum_{I_1=0}^{1} \dots \sum_{I_N=0}^{1} p(I_1, ..., I_N, t) \left(c_1^+\right)^{I_1} \dots \left(c_N^+\right)^{I_N} |0\rangle \\ &=: \sum_{\{I\}} p(\{I\}, t) \left(\prod_{i=1}^{N} \left(c_i^+\right)^{I_i} \right) |0\rangle \end{aligned} \tag{5.42}$$

and vacuum state $|0\rangle$. The creation and annihilation operators are defined by $c_i^+|0\rangle = |1\rangle$ and $c_i|1\rangle = |0\rangle$, and $\left(c_i^+\right)^2 |0\rangle = 0$ and $c_i|0\rangle = 0$; hence

$$c_i \;|I_i\rangle = I_i \cdot |1 - I_i\rangle \tag{5.43}$$

$$c_i^+ |I_i\rangle = (1 - I_i) \cdot |1 - I_i\rangle \tag{5.44}$$

and $\left(c_i^+\right)^2 |I_i\rangle = c_i^2 |I_i\rangle = 0$. We have anti-commutator rules on single lattice sites

$$[c_i, c_i^+]_+ := c_i c_i^+ + c_i^+ c_i = \mathbb{1} \tag{5.45}$$

and ordinary commutators for different lattice sites $i \neq j$

$$[c_i, c_j^+]_- := c_i c_j^+ - c_j^+ c_i = 0 \tag{5.46}$$

respectively

$$[c_i, c_j]_- = 0 \quad , \quad [c_i^+, c_j^+]_- = 0 \quad . \tag{5.47}$$

These are exactly the raising and lowering operators in [Brunel, Oerding and van Wijland (2000)] with

$$c^+ = \begin{pmatrix} 0 & 1 \\ 0 & 0 \end{pmatrix} \quad , \quad c = \begin{pmatrix} 0 & 0 \\ 1 & 0 \end{pmatrix} \quad , \tag{5.48}$$

for vectors

$$|1\rangle = \begin{pmatrix} 1 \\ 0 \end{pmatrix} \quad , \quad |0\rangle = \begin{pmatrix} 0 \\ 1 \end{pmatrix} \quad , \tag{5.49}$$

respectively product spaces of it for many particle systems as considered here (see Appendix D).

The authors of [Brunel, Oerding and van Wijland (2000)]) then use the Jordan–Wigner transformation to change to pure Fermi operators, with anti-commutation on single sites and on different sites, to obtain their path integrals. We use a different way.

The dynamics is expressed by

$$\frac{d}{dt}|\Psi(t)\rangle = \sum_{\{I\}} \left(\frac{d}{dt} p(\{I\}, t) \right) \prod_{i=1}^{N} \left(c_i^+\right)^{I_i} |0\rangle = L|\Psi(t)\rangle \tag{5.50}$$

where the master equation has to be inserted and evaluated to obtain the specific form of the operator L.

For the birth–death process Eq. (2.2) the Liouville operator requires some calculation

$$L = \sum_{i=1}^{N} (\mathbb{1} - c_i) \beta \left(\sum_{j=1}^{N} J_{ij} c_j^+ c_j \right) c_i^+ + \sum_{i=1}^{N} (\mathbb{1} - c_i^+) \alpha \, c_i \quad . \tag{5.51}$$

The term $(\mathbb{1} - c_i)$ guarantees the normalization of the master equation solution and $\beta J_{ij} c_j^+ c_j c_i^+$ creates one infected at site i from a neighbor j which is itself not altered. $c_j^+ c_j$ is simply the number operator on site j. Furthermore, αc_i removes a particle from site i, again ensuring normalization with $(\mathbb{1} - c_i^+)$.

Equation (5.70) and Eq. (5.72) are exactly the forms given in [Grassberger and de la Torre (1979)]. Hence they also use the raising and lowering operators.

However, it does not seem an easy task to construct the path integral from such a Liouville operator since no coherent states are constructed for the raising and lowering operators. Therefore, [Brunel, Oerding and van Wijland (2000)] proceed from these spin 1/2 operators to Fermi operators using Grassman variables for the coherent states, whereas [Cardy and Täuber (1998)] use Bose operators from the start, for which coherent states are easily available (e.g. [Le Bellac (1991)] and [Zinn-Justin (1989)]) hoping that rarely more than one particle will appear at a single site. However, [Park, Kim and Park (2000)] have emphasized once again the need for a rigoros formulation in terms of hard-core particles for which the exclusion principle on a single site and commutation on different sites is guaranteed.

This can be achieved by constructing δ-operators for bosons [van Wijland (2001)], as will be demonstrated for our birth–death process now.

5.3.3 *δ-Bosons for hard-core particles*

Defining Bose operators a^+ and a for states $|n\rangle$ with $n \in \mathbf{N}_0$ particles on one site by

$$a^+|n\rangle := |n+1\rangle \tag{5.52}$$

and

$$a|n\rangle := n \cdot |n-1\rangle \tag{5.53}$$

and the number operator $\hat{n} := a^+a$ with

$$a^+a|n\rangle = n|n\rangle \tag{5.54}$$

we can use δ-functions

$$\delta_{\hat{n},k}|m\rangle = \delta_{m,k}|m\rangle \tag{5.55}$$

with a suitable representation, e.g.

$$\delta_{\hat{n},k} = \frac{1}{2\pi}\int_{-\pi}^{\pi} e^{iu(\hat{n}-k)}\, du \tag{5.56}$$

[van Wijland (2001)]. $\delta_{m,k}$ is the ordinary Kronecker delta whereas $\delta_{\hat{n},k}$ is an operator defined by Eq. (5.55).

Thus we obtain for the birth–death process the following Liouville operator:

$$L = \sum_{i=1}^{N} (a_i^+ - \mathbb{1})\beta \left(\sum_{j=1}^{N} J_{ij} \delta_{\hat{n}_j,1} \right) \delta_{\hat{n}_i,0} + \sum_{i=1}^{N} (a_i - \mathbb{1})\alpha \, \delta_{\hat{n}_i,1} \quad . \tag{5.57}$$

which can be easily understood by replacing $\delta_{\hat{n}_i,1}$ in the bosonic theory by $c_i^+ c_i$ in the spin 1/2 theory, and $\delta_{\hat{n}_i,0}$ by $(\mathbb{1} - c_i^+ c_i)$ and simply replacing a_i by c_i and a_i^+ by c_i^+. Evaluating the resulting Liouville operator in terms of the spin 1/2 commutation rules then results exactly in Eq. (5.72) again.

5.3.4 *Path integral for hard-core particles in a birth–death process*

The path integral follows from integrating Eq. (5.70)

$$|\Psi(t)\rangle = \prod_{\nu=1}^{M} \left(1 + \Delta t \cdot L(t - \nu \cdot \Delta t)\right) |\Psi(t_M)\rangle \quad .$$

Hence with $\Delta t \to 0$, $M \to \infty$ and the finite time interval $\Delta t \cdot M = t - t_0$ we obtain for any expectation value $\langle f \rangle$ defined as

$$\langle f \rangle(t) := \sum_{\{n\}} f(\{n\}) p(\{n\}, t) = \langle P | f | \Psi(t) \rangle$$

with a Felderhof projection state $\langle P| := \langle 0 | e^{\sum_{i=1}^{N} a_i}$ [Felderhof (1971)] the path integral

$$\langle P | f | \Psi(t) \rangle = \int \dots \int \mathcal{D}\Phi_j^*(\tilde{t}) \mathcal{D}\Phi_j(\tilde{t}) f(t) \cdot e^{-\int_{t_0}^{t} d\tilde{t} \sum_{j=1}^{N} \left(\Phi_j^*(\tilde{t}) \frac{\partial \Phi_j(\tilde{t})}{\partial t} - \mathcal{L} \right)} \tag{5.58}$$

with

$$\mathcal{D}\Phi_j(\tilde{t}) := \prod_{j=1}^{N} \prod_{\nu=0}^{M} \frac{d\Phi_j(t_0 + \nu \cdot \Delta t)}{(2\pi i)^{N \cdot M}} \tag{5.59}$$

again in the limit $M \to \infty$ and $\Delta t \to 0$. The field variables $\Phi_j^*(\tilde{t})$ and $\Phi_j(\tilde{t})$ are introduced by coherent state integrals and replace the creation and annihilation operators by complex scalar variables. Here the Lagrange function is

$$\mathcal{L}(\Phi^*, \Phi) = \left(\beta \sum_{k=1}^{N} J_{jk} |\Phi_k|^2 e^{|\Phi_k|^2} - \alpha \Phi_j \right) \cdot \left(\Phi_j^* - 1 \right) e^{|\Phi_k|^2} \quad . \tag{5.60}$$

This can be compared to the path integrals used as a starting point for further analysis of directed percolation [Janssen (1981)] when we only use the lowest order of Φ in Taylor's expansion. Higher orders are expected to give irrelevant renormalization fields. Equation (5.58) has now exactly the form of a partition function. The path integral is thus ready for a further renormalization analysis (see [Cardy and Täuber (1998)]). On the numerical side, the real space renormalization as initially described by [Ma (1976)] for equilibrium systems makes steps toward further progress in understanding the spatial birth–death process near criticality. In Appendix E we give the derivations in more detail.

These calculations lead to critical exponents, but do not give any hint to where the critical threshold is. Furthermore, the exponents are rather crude approximations to the real values and often difficult to improve on due to the divergences in the approximation.

By far the best values as well for critical exponents as well as for the threshold values are obtained by series expansions. In the following section we will give a brief introduction to the ideas behind the series expansions and results for the non-linear birth and death processes respectively for the contact process, or our SIS epidemics. We will also show how to generalize the approach to more complicated models, e.g. the previously-studied SIRI epidemics.

5.4 Series Expansions

While mostly a renormalization analysis based on path integrals is performed on a Schrödinger-like equation from stochastic systems [Brunel, Oerding and van Wijland (2000); Park and Park (2005)], so that classical methods from particle physics can be used [Zinn-Justin (1989)], another method of series expansions based on a perturbation ansatz, as described in [Dickman and Jensen (1991); de Oliveira (2006)] is a useful way of obtaining phase separation points in epidemiological models like the SIS epidemics or phase separation lines in the SIRI model. Until now only simple systems like the one-dimensional contact process, i.e. the SIS epidemics, have been analyzed in this way.

5.4.1 *The SIS epidemic model revisited*

The Pauli matrices give lowering and raising operators as given in Eq. (5.48) for spin 1/2 systems (named after Wolfgang Pauli) [Brunel, Oerding and van Wijland (2000)]. The Pauli matrices are

$$\sigma_x := \begin{pmatrix} 0 & 1 \\ 1 & 0 \end{pmatrix} \quad , \quad \sigma_y := \begin{pmatrix} 0 & -i \\ i & 0 \end{pmatrix} \quad , \quad \sigma_z := \begin{pmatrix} 1 & 0 \\ 0 & -1 \end{pmatrix} \tag{5.61}$$

and from these the ladder operators are the raising operator

$$\sigma_+ := \frac{1}{2}\left(\sigma_x + i\sigma_y\right) = \begin{pmatrix} 0 & 1 \\ 0 & 0 \end{pmatrix} \tag{5.62}$$

and the lowering operator

$$\sigma_- := \frac{1}{2}\left(\sigma_x - i\sigma_y\right) = \begin{pmatrix} 0 & 0 \\ 1 & 0 \end{pmatrix} \quad . \tag{5.63}$$

Hence for the states

$$|1\rangle := \begin{pmatrix} 1 \\ 0 \end{pmatrix} \quad , \quad |0\rangle := \begin{pmatrix} 0 \\ 1 \end{pmatrix} \quad , \tag{5.64}$$

we have creation operator $c^+ = \sigma_+$ giving

$$c^+|0\rangle = |1\rangle \tag{5.65}$$

and annihilation operator $c = \sigma_-$ giving

$$c|1\rangle = |0\rangle \tag{5.66}$$

as they were previously in Eq. (5.48). Thus the SIS epidemic with master equation

$$\begin{aligned} \frac{d}{dt}p(I_1, ..., I_N, t) = &\sum_{i=1}^{N} w_{I_i,1-I_i}(t) \;\; p(I_1, ..., 1 - I_i, ..., I_N, t) \\ &- \sum_{i=1}^{N} w_{1-I_i,I_i}(t) \;\; p(I_1, ..., I_i, ..., I_N, t) \end{aligned} \tag{5.67}$$

for $I_i \in \{0, 1\}$ and transition rate

$$w_{I_i,1-I_i} = \beta \left(\sum_{j=1}^{N} J_{ij} I_j \right) \cdot I_i + \alpha \cdot (1 - I_i) \quad , \tag{5.68}$$

and

$$w_{1-I_i,I_i} = \beta \left(\sum_{j=1}^{N} J_{ij} I_j \right) \cdot (1 - I_i) + \alpha \cdot I_i \quad , \tag{5.69}$$

can be given in vector notation by

$$\frac{d}{dt} |\Psi(t)\rangle = L |\Psi(t)\rangle \tag{5.70}$$

for a Liouville operator L to be calculated from the master equation Eq. (5.67) and with state vector $|\Psi(t)\rangle$ defined by

$$\begin{aligned} |\Psi(t)\rangle &:= \sum_{I_1=0}^{1} \dots \sum_{I_N=0}^{1} p(I_1, ..., I_N, t) \left(c_1^+\right)^{I_1} \dots \left(c_N^+\right)^{I_N} |0\rangle \\ &=: \sum_{\{I\}} p(\{I\}, t) \left(\prod_{i=1}^{N} \left(c_i^+\right)^{I_i} \right) |0\rangle \end{aligned} \tag{5.71}$$

and vacuum state $|0\rangle$. L is given by

$$L = \sum_{i=1}^{N} (\mathbb{1} - c_i) \beta \left(\sum_{j=1}^{N} J_{ij} c_j^+ c_j \right) c_i^+ + \sum_{i=1}^{N} (\mathbb{1} - c_i^+) \alpha \, c_i \quad . \tag{5.72}$$

as derived previously, see Eq. (5.72).

5.4.2 *Perturbation analysis gives critical threshold*

The Liouville operator can be given in the form of a perurbation ansatz with an easily diagonalizable free operator W_0 acting only on single sites, and an interaction operator V contributing with strength λ to the interaction (for the one-dimensional contact process see e.g. [de Oliveira (2006)])

$$L = W_0 + \lambda \cdot V \tag{5.73}$$

with

$$W_0 := \sum_{i=1}^{N} \hat{B}_i \tag{5.74}$$

with $\hat{B}_i := (\mathbb{1} - c_i^+)\ c_i$ and without loss of generality $\alpha = 1$, and with

$$V := \sum_{i=1}^{N} \hat{Q}_i(\hat{n}_{i-1} + \hat{n}_{i+1}) \tag{5.75}$$

where $\hat{Q}_i := \frac{1}{Q}(\mathbb{1} - c_i)\ c_i^+$ with $Q = 2$ the number of neighbors in one dimension, and finally originating from the interaction term

$$\sum_{j=1}^{N} J_{ij} c_j^+ c_j = c_{i-1}^+ c_{i-1} + c_{i+1}^+ c_{i+1} \quad . \tag{5.76}$$

Identifying

$$\lambda := \beta \cdot Q \tag{5.77}$$

completes the perturbation ansatz Eq. (5.73) for the Liouville operator given in Eq. (5.72), as given in [de Oliveira (2006)]. With this ansatz the critical threshold and the critical exponents can be calculated very accurately via a scaling argument, e.g. for the time correlation function using the spectral gap [de Oliveira (2006); Martins, Aguiar, Pinto and Stollenwerk (2009)], and Padé approximation.

5.5 Generalization to the SIRI Epidemic Model

To generalize from the SIS epidemic model to more general cases like the SIRI, we need to extend the number of dimensions of the basic vectors for the states (see e.g. [Hinrichsen (2000)]), hence one more dimension for another particle type R besides I. For SIS we have

$$|0\rangle := \begin{pmatrix} 0 \\ 1 \end{pmatrix} =: |S\rangle \quad , \quad |1\rangle := \begin{pmatrix} 1 \\ 0 \end{pmatrix} =: |I\rangle \tag{5.78}$$

For the SIRI model we Therefore generalize to

$$|S\rangle := \begin{pmatrix} 0 \\ 0 \\ 1 \end{pmatrix} \quad , \quad |I\rangle := \begin{pmatrix} 0 \\ 1 \\ 0 \end{pmatrix} \quad , \quad |R\rangle := \begin{pmatrix} 1 \\ 0 \\ 0 \end{pmatrix} \quad . \tag{5.79}$$

Creation and annihilation operators are then given by

$$a^+|S\rangle = |I\rangle \quad , \quad a|I\rangle = |S\rangle \tag{5.80}$$

for infected, and for recovered we have

$$b^+|S\rangle = |R\rangle \quad , \quad b|R\rangle = |S\rangle \tag{5.81}$$

achieved by the following matrix representation

$$a^+ := \begin{pmatrix} 0\,0\,0 \\ 0\,0\,1 \\ 0\,0\,0 \end{pmatrix} \quad , \quad a := \begin{pmatrix} 0\,0\,0 \\ 0\,0\,0 \\ 0\,1\,0 \end{pmatrix} \tag{5.82}$$

and

$$b^+ := \begin{pmatrix} 0\,0\,1 \\ 0\,0\,0 \\ 0\,0\,0 \end{pmatrix} \quad , \quad b := \begin{pmatrix} 0\,0\,0 \\ 0\,0\,0 \\ 1\,0\,0 \end{pmatrix} \quad . \tag{5.83}$$

We can include the old two-dimensional creation and annihilation operators, now extended to three dimensions, as

$$c^+ := \begin{pmatrix} 0\,1\,0 \\ 0\,0\,0 \\ 0\,0\,0 \end{pmatrix} = b^+a \quad , \quad c := \begin{pmatrix} 0\,0\,0 \\ 1\,0\,0 \\ 0\,0\,0 \end{pmatrix} = a^+b \tag{5.84}$$

which can be expressed in terms of a^+, a and b^+, b. They give

$$c^+|I\rangle = |R\rangle \quad , \quad c|R\rangle = |I\rangle \quad . \tag{5.85}$$

These operators a^+, a, b^+, b and c^+, c are the ladder operators of the 8 Gell–Mann matrices λ_1, λ_2, ... , λ_8

$$a^+ = \frac{1}{2}\left(\lambda_6 + i\lambda_7\right) \quad , \quad a = \frac{1}{2}\left(\lambda_6 - i\lambda_7\right) \quad , \tag{5.86}$$

$$b^+ = \frac{1}{2}\left(\lambda_4 + i\lambda_5\right) \quad , \quad b = \frac{1}{2}\left(\lambda_4 - i\lambda_5\right) \quad , \tag{5.87}$$

$$c^+ = \frac{1}{2}\left(\lambda_1 + i\lambda_2\right) \quad , \quad c = \frac{1}{2}\left(\lambda_1 - i\lambda_2\right) \quad , \tag{5.88}$$

with the Gell-Mann matrices (named after Murray Gell-Mann, a generalization to the Pauli matrices that is well known in quantum chromodynamics

(QCD)) as representations of the SU(3) group (special unitary group of dimension 3)

$$\lambda_1 = \begin{pmatrix} 0 & 1 & 0 \\ 1 & 0 & 0 \\ 0 & 0 & 0 \end{pmatrix} \quad , \quad \lambda_2 = \begin{pmatrix} 0 & -i & 0 \\ i & 0 & 0 \\ 0 & 0 & 0 \end{pmatrix} \tag{5.89}$$

$$\lambda_4 = \begin{pmatrix} 0 & 0 & 1 \\ 0 & 0 & 0 \\ 1 & 0 & 0 \end{pmatrix} \quad , \quad \lambda_5 = \begin{pmatrix} 0 & 0 & -i \\ 0 & 0 & 0 \\ i & 0 & 0 \end{pmatrix} \tag{5.90}$$

$$\lambda_6 = \begin{pmatrix} 0 & 0 & 0 \\ 0 & 0 & 1 \\ 0 & 1 & 0 \end{pmatrix} \quad , \quad \lambda_7 = \begin{pmatrix} 0 & 0 & 0 \\ 0 & 0 & -i \\ 0 & i & 0 \end{pmatrix} \tag{5.91}$$

and

$$\lambda_3 = \begin{pmatrix} 1 & 0 & 0 \\ 0 & -1 & 0 \\ 0 & 0 & 0 \end{pmatrix} \quad , \quad \lambda_8 = \frac{1}{\sqrt{3}} \begin{pmatrix} 1 & 0 & 0 \\ 0 & 1 & 0 \\ 0 & 0 & -2 \end{pmatrix} \tag{5.92}$$

applied usually in QCD to the color states

$$|red\rangle = \begin{pmatrix} 1 \\ 0 \\ 0 \end{pmatrix} \quad , \quad |green\rangle = \begin{pmatrix} 0 \\ 1 \\ 0 \end{pmatrix} \quad , \quad |blue\rangle = \begin{pmatrix} 0 \\ 0 \\ 1 \end{pmatrix} \quad . \tag{5.93}$$

See e.g. [Hiesmayr (2006)] and [Koniorczyk and Janszky (2001)] for further elaboration.

For the commutator rules we now have

$$a\, a^+ = \mathbb{1} - (a^+ a + b^+ b) \quad , \tag{5.94}$$

and

$$b\, b^+ = \mathbb{1} - (a^+ a + b^+ b) \quad , \tag{5.95}$$

see [Park and Park (2005)].

5.5.1 *The SIRI epidemic model*

We consider the following transitions between host classes for N individuals being either susceptible S, infected I by a disease or recovered R

$$\begin{aligned} S+I &\xrightarrow{\beta} I+I \\ I &\xrightarrow{\gamma} R \\ R+I &\xrightarrow{\tilde{\beta}} I+I \\ R &\xrightarrow{\alpha} S \end{aligned}$$

resulting in the master equation [van Kampen (1992)] for variables S_i, I_i and $R_i \in \{0,1\}$, $i = 1,2,...,N$, for N individuals eventually on a regular grid, with constraint $S_i + I_i + R_i = 1$.

The first infection $S+I \xrightarrow{\beta} I+I$ occurs with infection rate β, whereas after recovery with rate γ the respective host becomes resistant up to a possible reinfection $R+I \xrightarrow{\tilde{\beta}} I+I$ with reinfection rate $\tilde{\beta}$. Hence the recovered are only partially immunized. For further analysis of possible stationary states we include a transition from recovered to susceptibles α, which might be simply due to demographic effects (or very slow waning immunity for some diseases). We will later consider the limit of vanishing or very small α. In case of demography that would be in the order of inverse 70 years, whereas for the basic epidemic processes like first infection β we would expect inverse a few weeks.

The master equation is explicitly given in the following form, as described in [Stollenwerk, Martins and Pinto (2007)]

$$\begin{aligned} \frac{d}{dt}\, p\,(S_1, I_1, R_1, S_2, I_2, R_2, ..., R_N, t) \\ = \sum_{i=1}^{N} \beta \left(\sum_{j=1}^{N} J_{ij} I_j \right) (1-S_i)\, p(S_1, I_1, R_1, ..., 1-S_i, 1-I_i, R_i ..., R_N, t) \\ + \sum_{i=1}^{N} \gamma (1-I_i)\, p(S_1, I_1, R_1, ..., S_i, 1-I_i, 1-R_i ..., R_N, t) \\ + \sum_{i=1}^{N} \tilde{\beta} \left(\sum_{j=1}^{N} J_{ij} I_j \right) (1-R_i)\, p(S_1, I_1, R_1, ..., S_i, 1-I_i, 1-R_i ..., R_N, t) \end{aligned} \qquad (5.96)$$

$$+\sum_{i=1}^{N} \alpha(1-R_i)\ p(S_1,I_1,R_1,...,1-S_i,I_i,1-R_i...,R_N,t)$$

$$-\sum_{i=1}^{N}\left[\beta\left(\sum_{j=1}^{N} J_{ij}I_j\right)S_i+\gamma I_i+\tilde{\beta}\left(\sum_{j=1}^{N} J_{ij}I_j\right)R_i+\alpha R_i\right]$$

$$\cdot p(...S_i,I_i,R_i...) \quad .$$

$J_{i,j} \in \{0,1\}$ are the elements of the $N \times N$ adjacency matrix J, symmetric and with zero diagonal elements. The formulation of the master equation is the same as that used in [Glauber (1963)].

5.5.2 *Transitions in the SIRI model*

From the rules above we can obtain the different parts of the Liouville operator corresponding to the transitions in the SIRI model (derived from the master equation transition rates).

Explicitly we have

$$S+I \xrightarrow{\beta} I+I \quad , \quad a^+|S\rangle = |I\rangle \tag{5.97}$$

meaning creation of an infected from a susceptible in interaction with another infected; hence

$$L_\beta = \beta\sum_{i=1}^{N}(\mathbb{1}-a_i)\left(\sum_{j=1}^{N} J_{ij}a_j^+a_j\right)a_i^+ \quad . \tag{5.98}$$

Next we have

$$R \xrightarrow{\alpha} S \quad , \quad b|R\rangle = |S\rangle \tag{5.99}$$

meaning creation of a susceptible from a recovered; hence

$$L_\alpha = \alpha\sum_{i=1}^{N}(\mathbb{1}-b_i^+)\ b_i \quad . \tag{5.100}$$

We then have

$$I \xrightarrow{\gamma} R \quad , \quad c^+|I\rangle = |R\rangle \tag{5.101}$$

meaning creation of a recovered from an infected; hence

$$L_\gamma = \gamma \sum_{i=1}^{N} (\mathbb{1} - c_i)\, c_i^+ = \gamma \sum_{i=1}^{N} (\mathbb{1} - a_i^+ b_i)\, b_i^+ a_i \tag{5.102}$$

Finally, we have

$$R + I \xrightarrow{\tilde{\beta}} I + I \quad , \quad c|R\rangle = |I\rangle \tag{5.103}$$

meaning creation of an infected from a recovered in interaction with another infected; hence

$$L_{\tilde{\beta}} = \tilde{\beta} \sum_{i=1}^{N} (\mathbb{1} - c_i^+) \left(\sum_{j=1}^{N} J_{ij} c_j c_j^+ \right) c_i = \tilde{\beta} \sum_{i=1}^{N} (\mathbb{1} - b_i^+ a_i) \left(\sum_{j=1}^{N} J_{ij} a_j^+ b_j b_j^+ a_j \right) a_i^+ b_i \tag{5.104}$$

With the commutation rules, especially Eq. (5.95), we obtain

$$c\, c^+ = a^+ b\, b^+ a = a^+ (\mathbb{1} - a^+ a - b^+ b) a = a^+ a \tag{5.105}$$

as the number operator for the infected, hence

$$L_{\tilde{\beta}} = \tilde{\beta} \sum_{i=1}^{N} (\mathbb{1} - b_i^+ a_i) \left(\sum_{j=1}^{N} J_{ij} a_j^+ a_j \right) a_i^+ b_i \tag{5.106}$$

The total Liouville operator is given by the sum of the individual Liouvilleans for the different transitions

$$L = L_\gamma + L_\alpha + L_\beta + L_{\tilde{\beta}} \tag{5.107}$$

where the first two terms $L_\gamma + L_\alpha$ are single-site or free contributions and the last two $L_\beta + L_{\tilde{\beta}}$ are interaction contributions to the Liouville operator, suitable for a perturbation analysis analoguous to ansatz Eq. (5.73).

We have formulated the stochastic spatial reinfection model SIRI in terms of creation and annihilation operators in a Schrödinger-like form [Stollenwerk and Aguiar (2008)], which opens the way for further analysis in terms of renormalization or series expansions via a perturbation ansatz to calculate critical thresholds and critical exponents. The creation and annihilation operators turn out to be the ladder operators of the Gell-Mann matrices of the SU(3) group representation [Stollenwerk and Aguiar (2008)].

These series expansions will, no doubt, give better results in the future than the pair approximations [Martins, Pinto and Stollenwerk (2009)], as it was previously shown for the pure SIS epidemics (Fig. 2.4).

Many more aspects on non-equilibrium systems from physics with occasional reference to epidemiology and ecology can be found in review articles from [Hinrichsen (2000)] and [Lübeck (2004)] and in the books of [Marro and Dickman (1999)] and [Tomé and de Oliveira (2001)].

In the following chapters we will give biological applications for the ideas and techniques described above. We are far from a detailed technical understanding of the critical behavior for complicated systems such as measles under vaccination, criticality in genetics and evolution towards criticality in bacterial meningitis. However, the mean field studies already indicate that the fluctuations in such biological systems can be understood as coming from system behavior close to a critical threshold. A more detailed analysis with the techniques described here will be very fruitful for future research.

Chapter 6

Criticality in Measles under Vaccination

6.1 Measles around Criticality

Measles epidemics in human populations have long been a subject of investigation. Studies on large data sets of country-wide notification numbers of cases since the beginning of the 1970s [London and Yorke (1973)] and [Yorke and London (1973)], lead to a classical model for childhood diseases [Dietz (1976)] including such aspects as a latent class beside the susceptible, infected and recovered and seasonality in the contact rate.

Due to the rather good empirical time series available, combined with various aspects of recent paradigmatic theories like deterministic chaos, there has been a vast literature about childhood diseases, namely measles, in prevaccination dynamics (see e.g. [Schwartz and Smith (1983); Schenzle (1984); Aron and Schwartz (1984); Schaffer (1985); Schaffer and Kott (1985)] to name just a few examples of the 1980s). The role of deterministic chaos in such measles notification data of the prevaccination era has been established since the 1990s (see e.g. [Olsen and Schaffer (1990); May and Sugihara (1990); Rand and Wilson (1991); Grenfell (1992); Bolker and Grenfell (1993); Stollenwerk (1992); Drepper, Engbert and Stollenwerk (1994)]).

Rather more recently, criticality in island populations has been investigated [Rhodes and Anderson (1996); Rhodes, Jensen and Anderson (1997)]. Though criticality is an older paradigm in physics than deterministic chaos, the advent of self-organized criticality in the 1990s has led more attention to aspects of criticality, namely power law behavior in epidemiological systems. Strangely, that classical criticality can play a role in epidemiology has only been discovered very recently. The critical threshold as an external parameter has to be tuned from the outside to reach the critical region. However, such a behavior seems to appear in vaccinated populations with decreasing vaccination levels, as we will show in this chapter.

We investigate a vaccinated population, where the only stable stationary state is the disease-free population and any invading disease cases lead to quickly extinct epidemics, when the vaccination level drops below the critical threshold where epidemics can take off. The consideration of dropping vaccination levels is motivated by the observation that in Britain, a discussion on side-effects of vaccines led to a dramatic drop in vaccine uptake [Jansen, Stollenwerk, Jensen, Ramsey, Edmunds, and Rhodes (2003)]. The mathematics to analyze the measles criticality (for a detailed description see [Jansen and Stollenwerk (2005)]) had previously been developed for the meningitis epidemiology which we will present in detail in Chapter 8. In this chapter we will describe the basic idea and show stochastic simulations around the vaccination threshold, which can be well compared with the knowledge about critical thresholds we have presented throughout this book.

6.2 The SIR Model

The basic SIR-model for a host population of size N divided into subclasses of susceptible, infected and recovered hosts [Anderson and May (1991)] is constructed as follows: with a rate α a resistent host becomes susceptible, or as a reaction scheme $R \xrightarrow{\alpha} S$. Then, susceptible meet infected with a transition rate β proportional to the number of infected. We divide by N to make the model scale invariant with population size, since we obtain a quadratic term in the variables as opposed to the linear term in the previous transition. As a reaction scheme we have $S + I \xrightarrow{\beta} I + I$. Finally, infected hosts can recover and become temporarily resistent with rate γ, hence $I \xrightarrow{\gamma} R$.[1]

The corresponding deterministic ordinary differential equation (ODE)

[1] We could also call this basic SIR-model an SIRS model, since transitions from R to S are allowed, but we will stick to SIR since later in an SIRYX model, parallel transitions prohibit a simple way of labeling. Hence, here SIR just means that we have three classes of hosts, S, I and R, to deal with as opposed to five classes in the more complicated model.

system reads

$$\frac{dS}{dt} = \alpha \cdot R - \beta \, \frac{I}{N} \cdot S$$
$$\frac{dI}{dt} = \beta \, \frac{I}{N} \cdot S - \gamma \cdot I \tag{6.1}$$
$$\frac{dR}{dt} = \gamma \cdot I - \alpha \cdot R$$

and describes merely the dynamics of the mean values for the total number of susceptibles, infected and recovered under the assumptions of mean field behavior and homogeneous mixing. Hence mean values of products can be replaced by products of means in the non-linear contact term $(\beta/N)\, I \cdot S$ (see Chapter 1). We include demographic stochasticity in the description of the epidemic. As such, for the basic SIR model we consider the dynamics of the probability $p(S, I, R, t)$ of the system to have S susceptibles, I infected and R recovered at time t, which is governed by a master equation (see [van Kampen (1992); Gardiner (1985)], and in a recent application to a plant epidemic model [Stollenwerk and Briggs (2000); Stollenwerk (2001)]. For state vectors $\underline{n}$, which is for the SIR model $\underline{n} = (S, I, R)$, the master equation reads

$$\frac{dp(\underline{n})}{dt} = \sum_{\underline{\tilde{n}} \neq \underline{n}} w_{\underline{n},\underline{\tilde{n}}} \; p(\underline{\tilde{n}}) - \sum_{\underline{\tilde{n}} \neq \underline{n}} w_{\underline{\tilde{n}},\underline{n}} \; p(\underline{n}) \tag{6.2}$$

with transition probabilities corresponding to those described above for the ODE system. The rates $w_{\underline{\tilde{n}},\underline{n}}$ are

$$\begin{aligned} w_{(S+1,I,R-1),(S,I,R)} &= \alpha \cdot R \\ w_{(S-1,I+1,R),(S,I,R)} &= \beta \cdot \frac{I}{N} \, S \\ w_{(S,I-1,R+1),(S,I,R)} &= \gamma \cdot I \end{aligned} \tag{6.3}$$

from which the rates $w_{\underline{n},\underline{\tilde{n}}}$ follow immediately as

$$\begin{aligned} w_{(S,I,R),(S-1,I,R+1)} &= \alpha \cdot (R+1) \\ w_{(S,I,R),(S+1,I-1,R)} &= \beta \cdot \frac{I-1}{N} \, (S+1) \\ w_{(S,I,R),(S,I+1,R-1)} &= \gamma \cdot (I+1) \quad . \end{aligned} \tag{6.4}$$

This formulation defines the stochastic process completely and will be the basis for modified models, e.g. terms for vaccination in the next section.

6.2.1 *The SIR model with vaccination*

Vaccination essentially transfers susceptible individuals directly into the recovered and now resistant class without passing through the infected stage. Hence only this transition with a vaccination transition rate, called v, has to be added to the basic model for measles. For simplicity, further details such as an exposed class and seasonality in the basic model are neglected here (however important they have been in studying the prevaccination era), since they would only have little impact in the present system. The basic mechanisms are already captured in the simple version presented here. The ODE system for the SIR model with vaccination reads

$$\begin{aligned}\dot{S} &= \mu(N-S) - \beta\frac{I}{N}\,S - v\cdot S \\ \dot{I} &= \beta\frac{S}{N}\,I - (\gamma+\mu)I \\ \dot{R} &= \gamma I - \mu R + vS\end{aligned} \tag{6.5}$$

with $v := \alpha\cdot\rho$ the vaccination rate. Here α is a time rate for the vaccination and ρ the proportion of vaccinated susceptibles. The stochastic version of this model will be shown below. We first analyze the mean field ODE system as given in Eq. (6.5) in more detail.

6.2.2 *Stationary states and vaccination threshold*

The stationary state solution already reveals a vaccination threshold. In the stationary state with $I^*=0$, hence no epidemics, we find

$$S_1^* = N\frac{\mu}{\mu+v} \quad , \quad I_1^* = 0 \quad , \quad R_1^* = N - S_1^* - I_1^* = N\frac{v}{\mu+v} \quad . \tag{6.6}$$

The conditions under which this solution is stable gives us a criterium for where the threshold we are looking for is located. The stability analysis looks at small perturbations of the ODE-system, Eq. (6.5), around the stationary state given by Eq. (6.6). Defining functions f, g and h from the ODE-system, Eq. (6.5),

$$\begin{aligned}\dot{S} &=: f(S,I,R) \\ \dot{I} &=: g(S,I,R) \\ \dot{R} &=: h(S,I,R) \quad ,\end{aligned} \tag{6.7}$$

we can evaluate the Jacobian matrix around the stationary state $\underline{x} := (S^*, I^*, R^*)$ as

$$\frac{d\underline{f}}{d\underline{x}} = \begin{pmatrix} \frac{\partial f}{\partial S} & \frac{\partial f}{\partial I} & \frac{\partial f}{\partial R} \\ \\ \frac{\partial g}{\partial S} & \frac{\partial g}{\partial I} & \frac{\partial g}{\partial R} \\ \\ \frac{\partial h}{\partial S} & \frac{\partial h}{\partial I} & \frac{\partial h}{\partial R} \end{pmatrix} \quad . \tag{6.8}$$

Hence for the disease-free stationary state we are looking at

$$\frac{d\underline{f}}{d\underline{x}} = \begin{pmatrix} -\mu - \beta\frac{I^*}{N} - v & -\beta\frac{S^*}{N} & 0 \\ \\ \beta\frac{I^*}{N} & \beta\frac{S^*}{N} - (\gamma + \mu) & 0 \\ \\ v & \gamma & -\mu \end{pmatrix} \quad . \tag{6.9}$$

The characteristic polynomial to find the eigenvalues of the Jacobian matrix is given by

$$\left(-\mu - \beta\frac{I^*}{N} - v - \lambda\right)\left(\beta\frac{S^*}{N} - (\gamma + \mu) - \lambda\right)(-\mu - \lambda) \tag{6.10}$$
$$+\beta\frac{I^*}{N} \cdot \beta\frac{S^*}{N} \cdot (-\mu - \lambda) = 0$$

As eigenvalues we have first $\lambda_3 = -\mu$, and, after some calculation, two non-trivial ones, namely $\lambda_2 = -(\mu + v)$ and

$$\lambda_1 = \beta\frac{S^*}{N} - (\gamma + \mu) \quad . \tag{6.11}$$

The requirement $\lambda_1 = 0$ gives the threshold value v_c, or critical vaccination value,

$$v_c = \frac{\mu}{\gamma + \mu}\Big(\beta - (\gamma + \mu)\Big) \quad . \tag{6.12}$$

This is the vaccination threshold we are looking for. Above the threshold there is only the disease-free state stable, and below it the disease can stay in a non-zero level. This also means that the disease-free state becomes unstable. A single infected imported from the outside can cause a major outbreak, leading the system to the now stable non-zero infection level. This is the picture one obtains from looking at the deterministic description of measles under vaccination, namely the ODE system, Eq. (6.5), and its analysis.

Below we will show that the stochastis system adds some qualitative features to this picture, namely that fluctuations already arise above the critical vaccination threshold, as an indication of proximity to that threshold. Exactly this information has been used by [Jansen, Stollenwerk, Jensen, Ramsey, Edmunds, and Rhodes (2003)] to demonstrate that in the UK the vaccination level is approaching the vaccination critical threshold coming from above. In the next subsection we analyze the non-zero infection stationary state, called the endemic state.

6.2.3 *Definition and expression for the reproduction number $\mathcal{R}$*

In the endemic stationary state $I^* \neq 0$ we find

$$S_2^* = N\frac{\gamma+\mu}{\beta} \quad , \quad I_2^* = N\left(\frac{\mu}{\beta}\left(\frac{\beta}{\gamma+\mu}-1\right)-\frac{v}{\beta}\right) \quad , \quad R_2^* = N - S_2^* - I_2^* \tag{6.13}$$

and with the heuristic definition of the reproduction level, $\mathcal{R}$, measured in stationarity

$$\mathcal{R}\cdot\frac{S_2^*}{N} = 1 \tag{6.14}$$

we obtain

$$\frac{S_2^*}{N} = \frac{1}{\mathcal{R}} = \frac{\gamma+\mu}{\beta} \quad . \tag{6.15}$$

Hence the critical vaccination threshold can be characterized by the reproduction number $\mathcal{R}$ in the following way

$$v_c = \frac{\mu}{\gamma+\mu}\Big(\beta-(\gamma+\mu)\Big) = \mu\mathcal{R}-\mu \quad . \tag{6.16}$$

6.2.4 *Vaccination level at criticality v_c*

At the criticality threshold v_c we obtain the classical results for the vaccination threshold [Anderson and May (1991)], namely $c_c = 1 - 1/\mathcal{R}$, where c_c is the critical value of the vaccination level c when writing the ODE for S in the form

$$\dot{S} = \mu((1-c)N-S) - \beta\frac{I}{N}\,S \tag{6.17}$$

as opposed to

$$\dot{S} = \mu(N - S) - \beta\frac{I}{N}\,S - v \cdot S \quad . \tag{6.18}$$

The argument is as follows: at criticality $v_c = \mu(\mathcal{R} - 1)$, and from the definition $\mathcal{R} = \beta/(\gamma + \mu)$, hence

$$\frac{S_c^*}{N} = \frac{\mu}{\mu + v_c} = \frac{1}{\mathcal{R}} \tag{6.19}$$

and

$$v_c \cdot S_c^* = \mu(\mathcal{R} - 1) \cdot \frac{N}{\mathcal{R}} = \mu\left(1 - \frac{1}{\mathcal{R}}\right) N \quad . \tag{6.20}$$

Thus, from

$$\dot{S} = \mu(N - S) - \beta\frac{I}{N}\,S - v \cdot S \tag{6.21}$$

we obtain in stationarity

$$\dot{S} = \mu(N - S) - \beta\frac{I}{N}\,S - v_c \cdot S_c^* \tag{6.22}$$

and hence

$$\dot{S} = \mu\left(\left(1 - \left(1 - \frac{1}{\mathcal{R}}\right)\right) N - S\right) - \beta\frac{I}{N}\,S \tag{6.23}$$

from which it follows directly that $c_c = 1 - 1/\mathcal{R}$.

6.2.5 *Realistic parameters for measles epidemics*

Rough estimates for measles parameters are: average life time $\mu^{-1} = 75$ years, average infection period $\gamma^{-1} = 0.02$ years from an estimate of around one week. The mean age of infection $(\beta \cdot I^*/N)^{-1} = 5$ years, with I^*/N in endemic equilibrium without vaccination, gives

$$\beta = (\gamma + \mu)\left(\frac{\mu^{-1}}{5\ \text{years}} + 1\right) \approx \gamma\frac{\mu^{-1}}{5\ \text{years}} = 750\ \text{years}^{-1} \quad . \tag{6.24}$$

The average age of vaccination can be $\alpha^{-1} \approx 1$ year to 3 years. Since only the percentage of to-be-vaccinated sucseptibles ρ varies, we do not have to specify this parameter very accurately, taking $\alpha^{-1} = 3$ years.

6.3 Stochastic Simulations

Simulations are done in the framework of master equations to capture the population noise using Gillespie's algorithm [Gillespie (1976); Gillespie (1978); Feistel (1977)]. The Gillespie algorithm, often also called minimal process algorithm, is a Monte Carlo method, in which after an event, i.e. a transition from state $\underline{n}$ to another state $\underline{\tilde{n}}$, the exponential waiting time is calculated as a random variable from the sum of all transition rates, after which the next transition is chosen randomly from all now possible transitions, according to their relative transition rates.

Hence, for the SIR-model we consider the dynamics of the probability $p(S,I,R,t)$, which is governed by a master equation (see [Gardiner (1985)] and [van Kampen (1992)]). For state vectors $\underline{n}$, for the SIR-model $\underline{n} = (S,I,R)$, the master equation reads

$$\frac{dp(\underline{n})}{dt} = \sum_{\underline{\tilde{n}} \neq \underline{n}} w_{\underline{n},\underline{\tilde{n}}} \; p(\underline{\tilde{n}}) - \sum_{\underline{\tilde{n}} \neq \underline{n}} w_{\underline{\tilde{n}},\underline{n}} \; p(\underline{n}) \tag{6.25}$$

with transition probabilities corresponding to those described above for the ODE system, here the rates $w_{\underline{\tilde{n}},\underline{n}}$ are

$$\begin{aligned}
w_{(S-1,I+1,R),(S,I,R)} &= \beta \cdot \frac{I}{N} \, S \\
w_{(S,I-1,R+1),(S,I,R)} &= \gamma \cdot I \\
w_{(S+1,I,R-1),(S,I,R)} &= \mu \cdot R \\
w_{(S+1,I-1,R),(S,I,R)} &= \mu \cdot I \\
w_{(S-1+1,I,R),(S,I,R)} &= \mu \cdot S \\
w_{(S-1,I,R+1),(S,I,R)} &= v \cdot S
\end{aligned} \tag{6.26}$$

from which the rates $w_{\underline{n},\underline{\tilde{n}}}$ follow immediately as

$$\begin{aligned}
w_{(S,I,R),(S+1,I-1,R)} &= \beta \cdot \frac{I-1}{N} \, (S+1) \\
w_{(S,I,R),(S,I+1,R-1)} &= \gamma \cdot (I+1) \\
w_{(S,I,R),(S-1,I,R+1)} &= \mu \cdot (R+1) \\
w_{(S,I,R),(S-1,I+1,R)} &= \mu \cdot (I+1) \\
w_{(S,I,R),(S+1-1,I,R)} &= \mu \cdot S \\
w_{(S,I,R),(S+1,I,R-1)} &= v \cdot (S+1) \quad .
\end{aligned} \tag{6.27}$$

6.3.1 *Stochastic bifurcation diagram for vaccine uptake*

We plot for each value for the vaccine uptake c the size of several epidemics after 3 years, when starting with one infected at the starting time. This shows that for high uptake rates only small epidmics are found, but for low values either the epidemics take off with high epidemic levels or still die out quickly (see Fig. 6.1). Large fluctuations are visible around the deterministic threshold value for c.

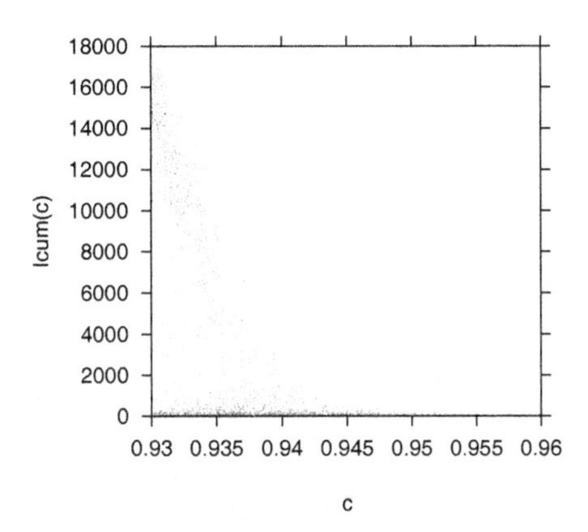

Fig. 6.1 Stochastic bifurcation diagram for vaccine uptake c.

At the equilibrium without infected (see above), we have for the susceptible hosts

$$S_1^* = N \cdot \frac{\mu}{\mu + v} \tag{6.28}$$

or in terms of c instead of v

$$S_1^* = N \cdot (1 - c) \quad . \tag{6.29}$$

Hence

$$v(c) = \frac{\mu c}{1 - c} \tag{6.30}$$

or

$$c(v) = \frac{v}{\mu + v} \quad . \tag{6.31}$$

In Fig. 6.1 a) we show stochastic simulations for various values of c, recalculating $v(c)$ for the simulations and starting each in the stationary values for S, R and one infected $I = 1$. The simulations are done for 3 years of epidemics. This summarizes the previous plots.

6.3.2 *Large outbreaks during decreasing vaccine uptake*

We consider the size of epidemics when lowering the uptake from 96% to 80%, introducing one infected at time t_i.

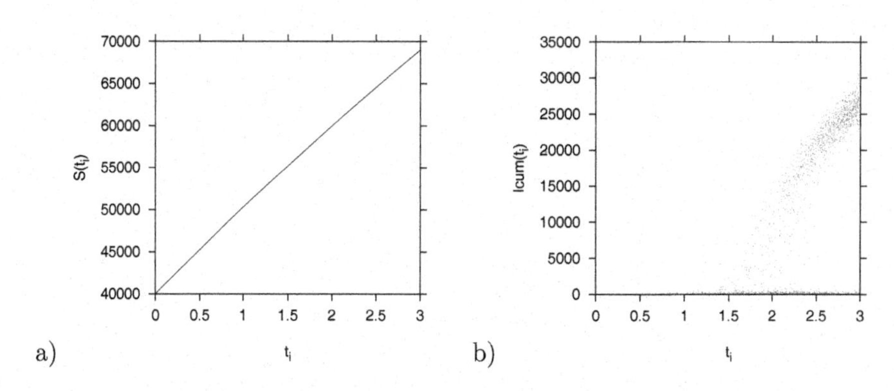

Fig. 6.2 a) $S(t)$ with $c_1 = 96\%$ and $c_2 = 80\%$ starting at $S(t_0) := S_1^*(c_1) = N(1 - c_1)$, but with c_2 (respectively $v(c_2)$) all the time. b) Size of epidemics when decreasing the uptake from 96% to 80%, introducing one infected at time t_i.

From the ODE for the susceptible class under vaccination

$$\dot{S} = \mu((1 - c)N - S) - \beta \frac{I}{N} S \tag{6.32}$$

with $c := c_2$ and $I(t) = 0$, hence no infected in the system, we obtain with $S(t_0) := S_1^*(c_1) = N(1 - c_1)$

$$S(t) = N(1 - c_1) + N(c_2 - c_1) \cdot (1 - e^{-\mu(t - t_0)}) \tag{6.33}$$

i.e. $S(t_\infty) = N(1 - c_2) = S_1^*(c_2)$ and $N(c_2 - c_1) = S(t_\infty) - S(t_0)$. For Fig. 6.2 b) we take $S(t) = S(t_i)$ as starting conditions for a stochastic simulation for one year of epidemics, introducing exactly one infected at time t_i into the system.

For the stochastic simulations and $S(t)$ we have to consider the dynamics of the fast vaccination time scale with $v(c)$ instead of c itself. So we start

with

$$\dot{S} = \mu(N - S) - vS \tag{6.34}$$

giving

$$S(t) = S(t_0) + (S(t_\infty) - S(t_0)) \cdot (1 - e^{-(\mu+v_2)(t-t_0)}) \tag{6.35}$$

with $S(t_0) = N(1-c_1) = N\mu/(\mu+v_1)$ and $S(t_\infty) = N(1-c_2) = N\mu/(\mu+v_2)$ equally if expressed in v or c. This results in the faster time scale for $S(t)$ with $(\mu + v_2)$ in the exponential instead of the slow μ only.

In summary, this shows that the decrease in vaccine uptake to low levels shows an effect only after some time, during which the number of susceptibles is built up and large epidemics are becoming more and more likely. Translated into the situation in the UK, large outbreaks of measles are to be expected after some years, since the vaccination level, varying regionally, has dropped from around 96% to as low as 85% and in some parts of London even below 80%.

A further analysis in terms of branching processes, including real-world data and its analysis, has been performed [Jansen and Stollenwerk (2005)] with the mathematical tools which we will present in Chapter 8 in application to the epidemiology of meningococcal dieseases. Hence we will not go into further detail here.

Chapter 7

Genetics and Criticality

7.1 Introduction

The field of population genetics studies the genetic make up of populations and the changes therein. The genomes of all living organisms contain large amounts of genes. Genes are the units of information that code certain traits, e.g. the color of one's eyes or whether or not one is resistant to a certain pesticide. Whenever organisms reproduce, they pass all, or a part of, their genes on to their progeny. From one generation to the next, the population of genes can change. It is this change in the genetic make up that is the main focus of genetics.

Not all individuals have identical genes. There is often variation between individuals in the information contained in a certain gene, because a single gene can be found in various difference forms. For instance, a gene for eye color might come in several different forms which could give the bearer brown or blue eyes. The different forms a gene can have are called alleles. Most higher organisms are diploid, meaning they have two copies of their genome, and therefore two copies of most of their genes (for some genes there is only one copy, like those involved in determination of the gender). If an organism has the same allele for both genes, it is said to be homozygous for this gene, whereas if it has two different copies, it is said to be heterozygous.

Allele frequencies in a population change through several processes. There are three causes that are of interest here. Firstly, occasionally genes are copied incorrectly and a mutation can be introduced in the population. Although this happens very infrequently, it is important because this can create novel genetic material and as such, mutations are the source of all genetic variation. Secondly, as not all parents have the same number of offspring, not all genes are copied into the next generation in identical numbers simply because of the sampling variation of the genes. This process is called random genetic drift and comprises the change in allele frequency through demographic stochasticity. Thirdly, the number of offspring an individual has can depend on the particular alleles it carries. For instance, in an environment where pesticides are used, an insect carrying an allele that gives it resistance against a pesticide could conceivable copy more of its genes into the next generation than an individual not carrying this allele. This is called selection and it will adapt a population to its environment.

Genetics studies the interplay between random processes, such as genetic drift, and deterministic forces, such as selection. If populations are large and there is strong selection, determinism is likely to prevail. On the other hand, if populations are small and selection is weak, the genetical change in a population will be mainly due to stochasticity. Within the genetics literature there has been a long-standing debate over which of these two views is most likely to explain the process of evolution. On the one hand, selectionists would claim that selective determinism explains most of the process of evolution. On the other hand, neutralists claim that most of the variation we see in natural populations is due to stochastic effects caused by neutral or weak selection. This is because most mutations have a negative effect on the selection of a gene and selection mostly is weak or absent.

However, these views are not mutually exclusive. It is most likely that in some case the selectionist view might hold sway, whereas in other cases the neutral view is more practical. As an illustration of this, the theory of neutral evolution, as developed by [Crow and Kimura (1970)], should be mentioned. This theory of neutral alleles forms the basis of the dating of evolutionary events through the use of the so-called molecular clock. This technique was an enormous technical advantage and has changed the way in which evolutionary biologists study the world.

Although the neutralist–selectionist debate is not as fiercely discussed as several decades ago, and both sides have accepted that the neutral theory is in a number of cases appropriate and in other cases can serve as a useful null model, the question remains to what extent stochasticity and determinism contribute to the evolutionary process, especially when selection is weak and populations are small. In such cases we can expect the dynamics to be critical or near-critical and we will apply the ideas of critical theory to some well-studied areas of genetics.

7.2 Models in Genetics

The most commonly used model in the context of genetics is arguably the Wright–Fisher model. Developed in the first decades of the twentieth century, this model his since formed a cornerstone of thinking in genetics. The model assumes that there is a population of N individuals which mix randomly and reproduce in non-overlapping generations. The total population size is constant and the number of offspring produced per parent follows a binomial distribution. To form the next generation the gametes are sampled randomly, and with replacement from the gamete pool. The Wright–Fisher model is popular because it describes a diploid population. It is to an extent mathematically tractable but many of the calculations follow straightforward, but somewhat tedious, calculations many of which are given in textbooks on genetics.

A second model that is frequently used is the Moran model [Moran (1958)]. This model also assumes a population of a fixed size, but has overlapping generations. Individuals are removed from this populations one by one, representing a process of death, and upon death are immediately replaced by new individuals which are sampled from the offspring of all individuals in the population. The Moran model assumes haploidy and is for that reason somewhat less popular. Because it is set in continuous time it tends to be mathematically easier, but in most cases there is a qualitative agreement between the results derived from the Moran model and the Wright–Fisher model.

In what follows, we will first give a mathematical description of the Moran model and relate it to some of the formalisms we have provided previously. We will then study the probability that a single mutation that arises in a population goes to fixation under various different scenarios, and demonstrate that these fixation probabilities obey scaling laws. These scaling laws allow us to distinguish between weak and strong selection, and as such make an interesting contribution to the selectionist–neutralist debate. In order to do so we, at times, have to make certain assumptions that are not quite in agreement with the biology. Although we do this for mathematical simplicity, we think this is justified because identical results can be derived for the Wright–Fisher model for which these assumptions are not made.

7.2.1 *The Moran model*

In this section we will formally describe the Moran model. Consider a population of N individuals, which hence have $2N$ copies of a gene, x of which are allele a and the remaining $2N - x$ are allele b. The total population size is assumed constant, and once a gene copy is lost through death, it is immediately replaced with a copy from the population of gametes. Note that at this point we make the simplifying assumption that gene copies die independently. This is of course not correct, but is an assumption of convenience. An artifact of this assumption is that the speed of the selective process is reduced by a factor 2. We will correct for this by multiplying the selection with a factor 2.

We will assume the the population is diploid and at all times in Hardy–Weinberg equilibrium, meaning the gametes are randomly paired to form diploid individuals. Hence the frequency of aa is $x^2/(2N)^2$, ab has frequency $2x(2N - x)/(2N)^2$ and bb has frequency $(2N - x)^2/(2N)^2$. The fitness (which in this case simply means the number of gametes contributed to the gamete pool) of the aa genotype is $1 + 2s$; the bb homozygote has a fitness which is set to unity, and the heterozygote ab has a fitness $1 + 2hs$. The total size of the gamete pool is

$$\bar{w} = \frac{x^2(1 + 2s) + 2x(2N - x)(1 + 2hs) + (2N - x)^2}{(2N)^2} \tag{7.1}$$

Let the death rate be μ. We will, for simplicity, assume that gametes are removed one by one (rather than in pairs as should really be the case for diploid individuals.) Following the death of an allele it will randomly replaced by an allele from the gamete pool. The replacement rates in the Moran model are

event	transition	rate
a replaced by a	$x \to x$	$\frac{2\mu x^2}{\bar{w}}(x(1+2s) + (2N - x)(1 + 2hs))$
a replaced by b	$x \to x - 1$	$\frac{2\mu x(2N-x)}{\bar{w}}(x(1 + 2hs) + (2N - x)))$
b replaced by a	$x \to x + 1$	$\frac{2\mu x(2N-x)}{\bar{w}}(x(1 + 2s) + (2N - x)(1 + 2hs))$
b replaced by b	$x \to x$	$\frac{2\mu(2N-x)^2}{\bar{w}}(x(1 + 2hs) + (2N - x))$

Or, if we define

$$d(x) = \frac{4\mu x(2N - x)}{\bar{w}}(hsx + N) \tag{7.2}$$

and

$$b(x) = \frac{4\mu x(2N - x)}{\bar{w}} \left(s(1-h)x + N(1+2hs)\right) \tag{7.3}$$

we can rewrite this in simplified form as

transition	rate
$x \to x+1$	$b(x)$
$x \to x-1$	$d(x)$
$x \to x$	$1 - b(c) - d(x)$

We can now write the master equation of this process as

$$\frac{\mathrm{d}p_x(t)}{\mathrm{d}t} = b(x-1)p_{x-1}(t) + d(x+1)p_{x+1}(t) \tag{7.4}$$

with the states $x = 0$ and $x = 2N$ acting as absorbing boundaries and hence we will set their densities to 0:

$$p_0(t) = 0$$

$$p_{2N}(t) = 0.$$

In matrix notation, we can write this as

$$\frac{\mathrm{d}\underline{p}}{\mathrm{d}t} = A\underline{p}(t) \tag{7.5}$$

where $\underline{p}(t)$ is the vector that contains the probabilities given by $(p_1(t), \ldots, p_{2N-1}(t))^{tr}$ and the matrix A is a tri-diagonal matrix that contains the transition rates as its non-zero elements:

$$A_{ij} = \begin{cases} 1 - b(i) - d(i) & \text{if} \quad i = j \\ b(i) & \text{if} \quad i = j+1 \\ d(j) & \text{if} \quad i = j-1 \\ 0 & \text{otherwise} \end{cases} \tag{7.6}$$

7.2.2 *The probability of fixation*

Although we did not keep track of the explicit probability of the system being in states 0 and $2N$, we can easily work out the probability of fixation of the two alleles. This probability increases over time with rate

$$b(2N-1)p_{2N-1}(t) \quad , \tag{7.7}$$

Rewritten in matrix notation this is

$$b(2N-1)e_{2N-1}^T \underline{p}(t)$$

where e_i is a column vector which entries are all zero except for the ith entry, which is 1. We can find the total probability of fixation by integrating this equation forward and taking the limit of t to infinity. This gives

$$b(2N-1)e_{2N-1}^T \int_0^\infty \underline{p}(t)\mathrm{d}t$$

We can solve Eq (7.5) to find $\underline{p}(t) = \exp(At)\underline{p}(0)$ and hence the probability of fixation is

$$b(2N-1)e_{2N-1}^T \left[A^{-1}\exp(At)\underline{p}(0)\right]_0^\infty = -b(2N-1)e_{2N-1}^T A^{-1}\underline{p}(0) \quad (7.8)$$

(This is because the vector $\underline{p}$ contains probabilities and hence all entries in the vector take values between 0 and 1. If absorption in states 2 or $2N$ eventually will take place, it is intuitively obvious that $\lim_{t\to\infty}\exp(At) = 0$.) If we start with a y copies of allele a in this population we have $\underline{p}(0) = e_y$ and we find that the probability that allele a goes to fixation, starting from y copies is

$$u(y) = -(A^{-1})_{2N-1,y} \quad . \quad (7.9)$$

There are analytical and numerical methods to calculate the inverse of a tri-diagonal matrix (see [Usmani (1994)] for analytical results or textbooks on numerical methods for the calculation of the inverse of a tri-diagonal matrix). The probability of allele a eventually becoming fixated in the population if it starts from a y copies is a well-known result given by

$$u(y) = \frac{1 + \sum_{j=1}^{y-1}\prod_{x=1}^{j}\frac{d(x)}{b(x)}}{1 + \sum_{j=1}^{2N-1}\prod_{x=1}^{j}\frac{d(x)}{b(x)}} \quad (7.10)$$

[Goel and Richter-Dyn (1974); Karlin and Taylor (1975)].

Although it is possible to express the fixation probability using special functions, this does not lead to transparent results. We will therefore discuss some cases of special interest.

Mutations are rare events. To evaluate the chance of a mutation establishing itself and taking over the population we will concentrate on the probability of fixation following the introduction of a single allele $u(1)$. An obvious first question is what the probability of fixation is in very large populations, in which the selection process is likely to dominate. If selection is negative, the allele will never establish itself in a large population. If there is selection for the allele, the probability of fixation in a large population is rouhgly given by $2hs$. In case of a recessive allele it cannot establish itself in a large population.

This can be shown as follows. If $h \neq 0$ and $s \neq 0$ we find

$$u(1) = \left(1 + \sum_{j=1}^{2N-1} \prod_{x=1}^{j} \frac{d(x)}{b(x)} \right)^{-1}$$

$$= \left(1 + \sum_{j=1}^{2N-1} \left(\frac{h}{1-h} \right)^{j} \frac{\Gamma(N\frac{1}{hs} + 1 + j)\Gamma(N\frac{1+2hs}{s(1-h)} + 1)}{\Gamma(N\frac{1+2hs}{s(1-h)} + 1 + j)\Gamma(N\frac{1}{hs} + 1)} \right)^{-1} . \quad (7.11)$$

Using $\Gamma(z+b) \approx z^{z+b} \exp(-z)\sqrt{2\pi z}$ for $z > 0$ we can approximate this for large N by

$$u(1) \approx \left(1 + \sum_{j=1}^{2N-1} \left(\frac{1}{1+2hs} \right)^{j} \right)^{-1} = \frac{2hs}{1+2hs} \left(1 - (1+2hs)^{-2N} \right)^{-1} . \quad (7.12)$$

In the limit of N tending to infinity this goes to

$$\lim_{N \to \infty} u(1) = \begin{cases} 0 & \text{if } hs \leq 0 \vee s < 0 \\ \frac{2hs}{1+2hs} & \text{if } hs > 0 \wedge s > 0 \end{cases} \quad (7.13)$$

This procedure works fine in the limit of h tending to one, which is the case of complete dominance. For this particular case this result is similar to the result found by Haldane that in the Wright–Fisher model the probability of fixation is roughly twice the selection coefficient [Crow and Kimura (1970)]

For the case of a recessive allele ($h = 0$) this argument does not hold and if $s \neq 0$ we find

$$u(1) = \left(1 + \sum_{j=1}^{2N-1} \prod_{x=1}^{j} \frac{N}{sx+N} \right)^{-1} = \left(1 + \sum_{j=1}^{2N-1} (N/s)^{j} \frac{\Gamma(N/s+1)}{\Gamma(N/s+j+1)} \right)^{-1} \quad (7.14)$$

To find an approximation for the probability if the selection coefficient, s, is small first observe that if there is no selection, i.e. $s = 0$, we find

$$u\left(\frac{1}{2N}\right) = \left(1 + \sum_{j=1}^{2N-1} \prod_{x=1}^{j} \frac{2N}{2N}\right)^{-1} = \frac{1}{2N} \quad . \tag{7.15}$$

For small s we have approximately

$$\begin{aligned}
\ln\left(u\left(\frac{1}{2N}\right)\right) &\approx -\ln 2N + \frac{2s}{2N}\left(\sum_{j=1}^{2N-1}\sum_{x=1}^{j}\frac{2N - x(2h-1)}{2N}\right) \\
&= -\ln 2N + \frac{2s}{2N}\left(\sum_{j=1}^{2N-1}\frac{j(4N-(j+1)(2h-1))}{4N}\right) \\
&= -\ln 2N + \frac{2s}{2N}\left(\frac{N(2N-1)(6N-(2h-1)(1+2N))}{6N}\right)
\end{aligned} \tag{7.16}$$

which for large N is approximately

$$\ln\left(u\left(\frac{1}{2N}\right)\right) = -\ln 2N + \frac{2s}{3}2N(2-h) \tag{7.17}$$

so that

$$u\left(\frac{1}{2N}\right) \approx \frac{1}{2N}e^{\frac{2s}{3}2N(2-h)} \quad . \tag{7.18}$$

The probability of fixation is approximately given by power law with exponent -1 if the population is small. If the power law dominates we could say that the behavior is similar as when selection is neutral, and hence that stochastic effects dominate. Identical results can be derived for the Wright–Fisher model. However, an analysis of the Wright–Fisher model shows that for large N the fixation probability obeys a power law with exponent $-\frac{1}{2}$. This behavior is also observed for the Moran model, but the rather crude approximation used here does not bear this out.

If the population is large, the exponential term will start to dominate. If selection is positive, the probability of fixation will be substantially bigger compared to the neutral case. If selection is negative, the fixation probability will be substantially smaller. If we assign, rather arbitrarily, that if the deviation is larger than a factor e for positive s (or $1/e$ for negative s) we can define weak selection as the case when

$$s < (2N)^{-1}\frac{3}{2(2-h)}$$

we can now define selection. Interestingly, whether or not selection is weak or not depends on the size of the population, but also on the way the phenotype is expressed. If the heterozygote is intermediate ($h = 1/2$) we find the boundary of weak selection is $s = (2N)^-1$, a result that is akin to results for haploid populations. However, if the allele is recessive (or dominant) we find the boundary is given by $s = \frac{3}{4}(2N)^{-1}$ ($s = \frac{3}{2}(2N)^{-1}$).

Most of the results on fixation probabilities are well known, but they are rarely in presented in relation to population size. We think this provides a useful link with the theory on criticality. This is particularly useful as it provides a way to discriminate between weak and strong selection. We have shown here that this boundary does not only depend on population size, but also on details such as the way in which a gene inherits.

7.3 Mean Time until Fixation

Although the probability of fixation gives a useful indication of rate with which new mutations will go to fixation, in large populations this process can take a long time. To see how the mean time until fixation depends on population size, we again can derive results from the transition matrix. The mean time until fixation in state $2N$ as

$$b(2N-1)e_{2N-1}^T \int_0^\infty t\underline{p}(t)\,\mathrm{d}t \tag{7.19}$$

We can substitute $\underline{p}(t) = \exp(At)\underline{p}(0)$ and integrate by parts to find

$$b(2N-1)e_{2N-1}^T A^{-2}\underline{p}(0). \tag{7.20}$$

This provides an easy formalism for the numerical calculation of the times until fixation. It is possible to derive closed form expressions, but these do not normally lead to useful insights

A simple expression only has been obtained for the neutral case. [Crow and Kimura (1970)] and [Goel and Richter-Dyn (1974)] show that for the Wright–Fisher model, the number of generations until fixation starting from a single gene, if fixation is bound to occur, is $4N\frac{1-y}{y}|\ln(1-y)|$. For small y this is approximately equal to $4N$ and therefore in large diploid populations the average number of generations until fixation, starting from a single gene, is approximately twice the number of genes.

Chapter 8

Evolution to Criticality in Meningococcal Disease

Although meningitis and septicaemia are only rarely observed diseases and often clustered in smaller or larger epidemics, the bacteria causing the disease can be detected in as many as 30 or 40% of the host population as harmless commensals. Frequently, mutations in these bacteria occur and from time to time they make the severe mistake to harm their hosts seriously, in former times almost always fatally.

We model the host dynamics for meningitis and septicaemia as a simple SIR model for the harmless strain of bacteria with additional classes for the infection with mutant bacteria, called Y hosts, and severely diseased cases X. With this model we can show that huge fluctuations occur when the chance of a mutant causing a diseased case, called pathogenicity, is small. Furthermore, we can show that in systems with mutations of various values of pathogenicity only those with small pathogenicity are present for significant periods of time. For such small values of pathogenicity we can show power law behavior of the size distribution of epidemics (see [Stollenwerk and Jansen (2003,a)] for details). Thus we can demonstrate that the system is in criticality. Evolution towards criticality can be demonstrated by looking at a continuum of pathogenicities. The infection dynamics will quickly cause the extinction of any mutant infections with high pathogenicity, leaving the infections with low pathogenicity for a longer time in the system, so that they accumulate in the course of the process [Stollenwerk and Jansen (2003,b)]. data inspection corroborates this picture [Stollenwerk, Maiden and Jansen (2004)].

Meningitis data sets from England and Wales, Norway and the USA are analyzed. The SIRYX stochastic system close to criticality, a model for accidental pathogens, is adjusted to the seven years long weekly England and Wales data that aims to capture the fast response to seasonal forcing in the contact rate. On this level it cannot be distinguished from a simpler SIRX system as previously suggested.

Surprisingly in simulations, huge epidemic outbreaks are found lasting over decades as they are observed in the 40 years long yearly Norwegian data. Only the parameter set obtained from the short England and Wales data was used and no information on longer fluctuation was included. Considering these long term fluctuations, the SIRYX describes the data variance much better than the simpler SIRX system. The simple SIRX system can be rejected on the basis of showing effectively Poisson-like behavior, and the variance over mean ratio being close to unity, which clearly contradicts the data.

In data from the USA, though weaker, both, the seasonality from the short highly-resolved England and Wales data as well as the huge decade-long outbreaks observed in the long-term Norway data, are visible. Hence, critical fluctuations can explain parsimoniously different aspects of observed data, seasonality and decade-long fluctuations without postulating different mechanisms for either of them.

This section is based on previous work [Stollenwerk and Jansen (2003,a); Stollenwerk and Jansen (2003,b); Stollenwerk, Maiden and Jansen (2004)], but also includes later results and gives the context of this application to the theoretical concepts outlined in this book in the previous chapters.

8.1 Accidental Pathogens

In recent years we have seen different scientific disciplines working together fruitfully. A rather classical example from the beginning of the twentieth century is the explanation of chemical reactions by physical atomic models. Evolutionary biology and epidemiology, accompanied by statistical physics of critical phenomena, present a new picture to explain unpredicted outbreaks of a severe disease as we will show in a case study on meningococcal infection. This case study will also provide a new mechanism to understand the epidemiology of this particular example, meningococcal disease, better than previous models could, as it will serve as a test bed for general principles discussed in evolutionary biology, namely that minor errors on the individual level can cause greater harm on the population level than a major error being subject to strong selection.

The quantitative mechanism is provided by the theory of critical phenomena in statistical physics. The closer the system is to criticality, here the smaller the error of a pathogen is, the larger are the fluctuations that have to be expected.

It is the occurence of these large fluctuations – large outbreaks between long phases of minor infection levels – which has puzzled the epidemiology of meningococcal disease for a long time, accompanied by the large variability of the pathogen, the bacterium *Neisseria meningitidis*. Whereas traditional quantitative epidemiology concentrates on models including infected hosts as the once clinically visible, in meningococcal infections the majority of the infections is undetected, meaning that the bacteria live within the host, often going unnoticed. In the case of meningococci as many as 20% or more of the human population is infected without noticing it and never becoming clinically ill. In certain age classes even higher numbers have been reported. Only in a very small proportion of the infections the host develops the disease, but then in rapid and often life threatening form, leaving patients with missing limbs in the case of *septicaemia*, or severe brain damage in the case of *meningitis*.

The small probability of pathogenicity causes huge critical fluctuations on the population level, a mechanism most clearly visible in meningococcal disease, but possibly underlying many other epidemiological systems, not only of bacterial infections but also viral infections. Since bacteria have their own metabolism, they are able to live with their host little affected, whereas viruses have to hijack host cells in order to be able to reproduce. For example, in polio infection most of the time the viruses live in the host's gut undetected. It is only by entering nerve cells they cause the very severe disease, a rarely happening event.

As epidemiology is one of the best data sources of biological interactions, especially notifiable diseases, and micro-organisms in a hostile environment such as the pathogen-host interaction are the fastest mutating biological systems, this is the ideal set-up for evolutionary biology to be tested quantitatively. However, the implications might well be wider reaching, but less obviously detectable, for example in predator–prey systems due to small numbers of individuals and generally less observational effort.

On the technical side, the critical fluctuations that are so crucial in understanding major epidemic outbreaks were originally investigated in very large physical systems, in numbers going well beyond the world's human population. Although fingerprints of a critical state can be obtained near criticality, it becomes increasingly difficult to investigate critical quantities the closer to criticality the system is. So we can only hope to find these fingerprints, but not really attempt to measure accurately critical exponents, for example. It should also be mentioned that we are in a so-called non-equilibrium critical system, a birth–death process effectively, whereas the most powerful characterization of criticality is obtained in equilibrium systems, like the famous Ising model for magnetic phase transitions. However, as the system under investigation evolves on its own towards a critical state, we can expect the system to be reasonably close to criticality most of the time in order to detect the large fluctuations reliably in empirical data.

8.2 Modeling Infection with Accidental Pathogens

Classically, epidemics are modeled dividing the host population into susceptibles S, infected I and recovered R, where the infected are clinically detected, i.e. sick. A standard model is the SIR model [Anderson and May (1991)] as described earlier in Chapter 6.

However, in the epidemiology of meninoccocal disease it is well established that infection with the bacteria *Neisseria meningitidis* itself is not noticed by the host, with infection levels in a given population being as high as 10% in industrialized countries and reaching up to 40% in certain age groups, but only the much rarer cases, when the infection leads to diseases like meningitis or septicaemia are recognized by the host. [Coen, Cartwright and Stuart (2000)]. Technically, the bacteria live as commensals in the nasopharynx of the hosts as an unnoticed, completely harmless infection, now identified by harmless infected hosts I. Occasionally, the *Neisseria* bacteria cross the nasal wall into the blood stream and immediately cause severe damage to the host, identified as septicaemia, or transferred even further into the brain causing meningitis, where we label affected hosts by X, since without medication hosts often die or are left disabled, hence removed from normal social interaction. The resulting SIRX model would allow a transition from harmless infected to diseased hosts with a small probability rate ε. It is only the number of diseased cases X which is recorded in empirical data of notification of meningococcal infection. We will investigate the quantitative outcome of this SIRX model in respect of statistics of the disease cases X below, but can already say here that the Poisson-like behavior of the SIRX model does not even account for basic epidemiological findings that meningitis and septicaemia often appear in clusters with pronounced phases of silence between outbreaks.

Only when we include another finding of the biology of *Neisseria* bacteria in the modeling of its epidemiology can we obtain such clustered outbreaks, namely that the bacteria are highly mutating in order to escape the continuous attacks by the hosts' immune system during harmless carriage. The different mutants of the bacterium have different likelihoods of making the mistake to harm their hosts significantly by causing severe disease. Hence in the easiest modeling set-up where we found clustered outbreaks [Stollenwerk and Jansen (2003,a)] we distinguished between harmless infection never causing disease, named I class, and potentially harmful infection with a different mutant strain of the bacteria, named Y class, from which with a small probability rate ε, the pathogenicity, disease cases X are created. For pathogenicity close to its critical value of zero we found huge fluctuations, to be expected from the theory of critical phenomena in physics of condensed matter [Stanley (1971); Landau and Binder (2000)] and in biology of critical birth and death processes [Grassberger and de la Torre (1979); Grassberger (1983)] (for a general introduction see [Warden (2001)]). These fluctuations are giving rise to clustered outbreaks in disease cases X in our SIRYX model [Stollenwerk and Jansen (2003,a)].

8.2.1 *The meningitis model: SIRYX*

In order to describe the behavior of pathogenic strains added to the basic SIR system we include a new class Y of individuals infected with a potentially pathogenic strain. We will assume that such strains arise by e.g. point mutations or recombination through a mutation process with a rate μ in the scheme $S + I \xrightarrow{\mu} Y + I$. For symmetry we also allow the mutants to back-mutate with rate ν, hence $S + Y \xrightarrow{\nu} I + Y$.

The major point here in introducing the mutant is that the mutant has the same basic epidemiological parameters α, β and γ as the original strain and only differs in its additional transition to pathogenicity with rate ε.

These mutants cause disease with rate ε, which will turn out to be small later on, hence the reaction scheme is $S + Y \xrightarrow{\varepsilon} X + Y$. This sends susceptible hosts into an X class, which contains all hosts who develop the symptomatic disease. These are the cases wich are detectable as opposed to hosts in classes Y and I that are asymptomatic carriers who cannot be detected easily.

The state vector in the extended model, the SIRYX model, is now $\underline{n} = (S, I, R, Y, X)$. The mutation transition $S + I \xrightarrow{\mu} Y + I$ fixes the master equation transition rate $w_{(S-1,I,R,Y+1,X),(S,I,R,Y,X)} = \mu \cdot (I/N) \cdot S$. In order to denote the total contact rate still with the parameter β, we keep the balancing relation

$$w_{(S-1,I+1,R,Y,X),(S,I,R,Y,X)} + w_{(S-1,I,R,Y+1,X),(S,I,R,Y,X)} = \beta \cdot \frac{I}{N} \cdot S \quad (8.1)$$

and obtain for the ordinary infection of normal carriage the transition rate $w_{(S-1,I+1,R,Y,X),(S,I,R,Y,X)} = (\beta - \mu) \cdot (I/N) \cdot S$. To denote the total rate of contacts a susceptible host can make with any infected, either normal carriage I or mutant carriage Y, by β, we obey the balancing equation

$$\sum_{\underline{\tilde{m}} \neq \underline{m}} w_{(S-1,\underline{\tilde{m}}),(S,\underline{m})} = \beta \, \frac{I+Y}{N} \cdot S \quad (8.2)$$

for $\underline{m} = (I, R, Y, X)$. With the above-mentioned transitions this fixes the master equation rate $w_{(S-1,I,R,Y+1,X),(S,I,R,Y,X)} = (\beta - \nu - \varepsilon) \cdot (Y/N) \cdot S$.

For completeness, we introduce a recovery from the severe meningitis and septicaemia respectively with rate φ, hence $X \xrightarrow{\varphi} S$. With regard to meningitis and septicaemia, in many cases the disease is fatal, hence $\varphi = 0$. With medication the sufferers often survive, but are hospitalized for a long time and then may suffer from resulting impairments. So for the theoretical analysis we will still keep $\varphi = 0$, which might be changed when analyzing more realistic situations or recent data.

For the SIRYX-system the transition probabilities $w_{\underline{\tilde{n}},\underline{n}}$ are then given (omitting unchanged indices in $\underline{\tilde{n}}$, with respect to $\underline{n}$) by

$$\begin{array}{rll}
w_{(R-1,S+1),(R,S)} = \alpha \cdot R & , & R \xrightarrow{\alpha} S \\
w_{(S-1,I+1),(S,I)} = (\beta - \mu) \cdot \frac{I}{N} S & , & S + I \xrightarrow{\beta-\mu} I + I \\
w_{(S-1,Y+1),(S,Y)} = \mu \cdot \frac{I}{N} S & , & \xrightarrow{\mu} Y + I \\
w_{(I-1,R+1),(I,R)} = \gamma \cdot I & , & I \xrightarrow{\gamma} R \\
w_{(S-1,Y+1),(S,Y)} = (\beta - \nu - \varepsilon) \cdot \frac{Y}{N} S & , & S + Y \xrightarrow{\beta-\nu-\varepsilon} Y + Y \\
w_{(S-1,I+1),(S,I)} = \nu \cdot \frac{Y}{N} S & , & \xrightarrow{\nu} I + Y \\
w_{(S-1,X+1),(S,X)} = \varepsilon \cdot \frac{Y}{N} S & , & \xrightarrow{\varepsilon} X + Y \\
w_{(Y-1,R+1),(Y,R)} = \gamma \cdot Y & , & Y \xrightarrow{\gamma} R \\
w_{(X-1,S+1),(X,S)} = \varphi \cdot X & , & X \xrightarrow{\varphi} S
\end{array} \quad (8.3)$$

along with the respective reaction schemes. Again from $w_{\underline{\tilde{n}},\underline{n}}$ the rates $w_{\underline{n},\underline{\tilde{n}}}$ follow immediately. This defines the master equation for the full SIRYX system.

8.2.2 *The invasion dynamics of mutant strains*

Before we proceed with further theoretical analysis of the model we now demonstrate basic properties of our SIRYX model in simulations of the master equation, using the Gillespie algorithm, also known as minimal process algorithm [Gillespie (1976)]. This is a Monte Carlo method, in which after an event, i.e. a transition from state $\underline{n}$ to another state $\underline{\tilde{n}}$, the exponential waiting time is calculated as a random variable from the sum of all transition rates, after which the next transition is chosen randomly from all now possible transitions according to their relative transition rates.

To investigate the dynamics of the infection with mutants, class Y, in relation to the normal carriage I with harmless strains, we first fix the basic SIR subsystem's parameters to the values $\alpha := 0.1$, $\beta := 0.2$ and $\gamma := 0.1$.

The endemic equilibrium of the SIR system is given by

$$S^* = N\frac{\gamma}{\beta} \quad , \qquad I^* = N\frac{\alpha}{\beta}\left(\frac{\beta-\gamma}{\alpha+\gamma}\right) \quad , \qquad R^* = N - S^* - I^* \qquad (8.4)$$

as can be seen from Eq. (6.1) setting the left-hand side of each subequation to zero and $\gamma := 0.1$. This equilibrium would correspond to labelling 2, hence S_2^* etc., in previous chapters. As to the parameters used, we find in equilibrium a normal level of carriage of harmless infection of about 25% in our total population of size N. This is in agreement with reported levels of carriage for *Neisseria meningitidis*. Average duration of carriage is in the order of 10 months, hence we choose $\gamma = 0.1$. We assume the duration of immunity to be the same as the duration of carriage. In equilibrium this results in the ratio of $S^* : I^* : R^* = 2 : 1 : 1$. However, the qualitative results are not affected by these first guesses of parameter values, but rather the order of magnitude.

Interesting behavior is observed if the pathogenicity ε is too large for the hyperinvasive strain to take over, but small enough to create large outbreaks of mutant infected hosts Y before becoming extinct again. In Fig. 8.1 we show two simulations in this ε-region, first $\varepsilon = 0.05$ in Fig. 8.1a) and 8.1c), then a ten times smaller ε in Fig. 8.1b) and 8.1d). For high pathogenicity ε we find relatively low levels of mutants Y, fewer than 20 cases in Fig. 8.1c), and at the end of the simulation roughly between 15 and 80 hospital cases X, Fig. 8.1a). For smaller pathogenicity ε, Fig. 8.1d), we find much larger fluctuations in the number of mutants Y with peaks of more than 80 mutant-infected hosts. Though the probability rate of causing disease ε is ten times smaller than in the previous simulation, we find at the end of this simulation similar numbers of disease cases X, Fig. 8.1b). We observe larger fluctuations and sometimes many more outbreaks of diseased cases although the probability of causing disease is smaller.

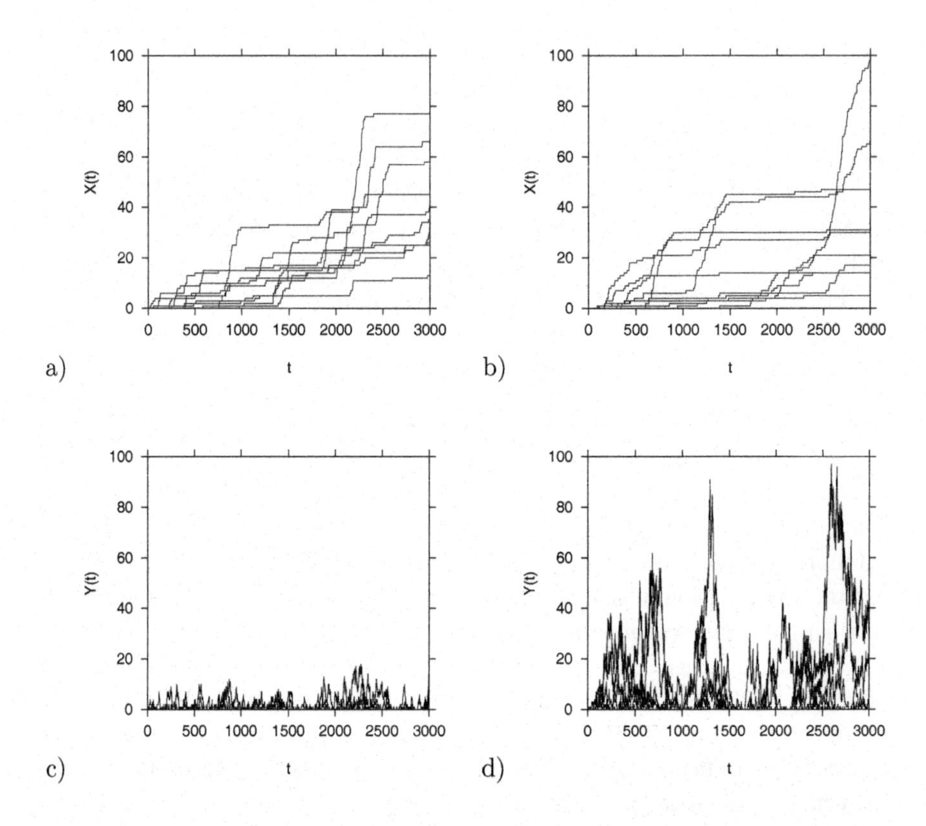

Fig. 8.1 For the SIRYX model we show simulations of 10 runs for two different values of pathogenicity ε. In a) the cumulative number of diseased cases X is shown for a certain ε, whereas in b) a value ten times smaller is used. Seemingly paradoxically, the cumulative number of diseased cases does not also decrease by a factor of ten, but fluctuates more wildly, sometimes leading to even higher numbers of diseased. The paradox is explained by inspecting the numbers of mutant infected in c) for the ε as in a), and with much higher values of mutant infected in d) for the ten times smaller ε, due to their smaller disadvantage compared to the harmless carriage.

This counter intuitive result can be understood by considering the dynamics of the hyperinvasive lineage in detail. We will do so by analyzing a simplified version of our SIRYX model analytically.

8.2.3 *Divergent fluctuations for vanishing pathogenicity*

For pathogenicity ε larger than the mutation rate μ the hyperinvasive lineage normally does not attain very high densities compared to the total population size. Therefore, we can consider the full system as being composed of a dominating SIR system which is not really affected by the rare Y and X cases, calling it the SIR heat bath, and our system of interest, namely the Y cases and their resulting pathogenic cases X, is considered to live in the SIR heat bath.

Taking into account Eq. (8.4) for the stationary values of the SIR system, we obtain for the transition rates (compare Eq. (8.34)) of the remaining YX-system

$$\begin{aligned}
w_{(S^*,Y+1),(S^*,Y)} &= \mu \cdot \tfrac{S^*}{N}\, I^* && =: c \\
w_{(S^*,Y+1),(S^*,Y)} &= (\beta - \nu - \varepsilon) \cdot \tfrac{S^*}{N}\, Y && =: b \cdot Y \\
w_{(S^*,X+1),(S^*,X)} &= \varepsilon \cdot \tfrac{S^*}{N}\, Y && =: g \cdot Y \\
w_{(Y-1,R^*),(Y,R^*)} &= \gamma \cdot Y && =: a \cdot Y \\
w_{(X-1,S^*),(X,S^*)} &= \varphi \cdot X \quad .
\end{aligned} \tag{8.5}$$

All terms not involving Y or X vanish from the master equation, since the gain and loss terms cancel each other out for such transitions. If we neglect the recovery of the disease cases to susceptibility, as is reasonable for meningitis, hence $\varphi = 0$, we are only left with Y-dependent transition rates. Hence for the YX-system we obtain the master equation

$$\begin{aligned}
\frac{d}{dt} p(Y,X,t) &= (b \cdot (Y-1) + c)\, p(Y-1,X,t) + a \cdot (Y+1)\, p(Y+1,X,t) \\
&\quad + g \cdot Y\, p(Y,X-1,t) \quad - (bY + aY + gY + c)\, p(Y,X,t) \quad .
\end{aligned} \tag{8.6}$$

This gives for the marginal distribution $p(Y,t) := \sum_{X=0}^{\infty} p(Y,X,t)$ the master equation for a simple birth–death process with birth rate $b := (\beta - \nu - \varepsilon) \cdot \frac{S^*}{N}$, death rate $a := \gamma$ and a migration rate $c := \mu \cdot \frac{S^*}{N}\, I^*$. In the definition of the marginal distribution we take the upper limit of the summation to infinity, since we assume numbers of X and Y cases to be well below the stationary values of the SIR system, i.e. they will not be affected by any finite upper boundary. We will check the validity of this assumption later with simulations of the full SIRYX system.

Hence we have

$$\frac{d}{dt}p(Y,t) = (b\cdot(Y-1)+c)\,p(Y-1,t) + a\cdot(Y+1)\,p(Y+1,t) \\ -(bY+aY+c)\,p(Y,t) \tag{8.7}$$

for $Y \in \mathbb{N}$ and as boundary equation, i.e. for $Y = 0$

$$\frac{d}{dt}p(Y=0,t) = a\cdot p(Y=1,t) - c\cdot p(Y=0,t) \quad . \tag{8.8}$$

For the ensemble mean $\langle Y\rangle := \sum_{Y=0}^{\infty} Y\cdot p(Y,t)$ we obtain, using the above master equation,

$$\frac{d}{dt}\langle Y\rangle = (b-a)\cdot\langle Y\rangle + c \quad . \tag{8.9}$$

For the variance, being defined as $Var(t) := \langle Y^2\rangle - \langle Y\rangle^2$, we obtain

$$\frac{d}{dt}\underbrace{\left(\langle Y^2\rangle - \langle Y\rangle^2\right)}_{=:Var(t)} = 2(b-a)\left(\langle Y^2\rangle - \langle Y\rangle^2\right) + (b+a)\cdot\langle Y\rangle + c \quad . \tag{8.10}$$

We can simplify further by neglecting the mutation and back-mutation terms, hence $c = 0$, and $\nu = 0$ in the definition for b, and solve the two ODEs for mean $Y(t) := \langle Y\rangle$ and variance $Var(t)$, noting that

$$b - a = (\beta-\varepsilon)\cdot\frac{S^*}{N} - \gamma \quad = -\varepsilon\cdot\frac{S^*}{N} \tag{8.11}$$

is proportional to ε. We set $g := \varepsilon\cdot\frac{S^*}{N}$. The ODEs then read

$$\dot{Y}(t) = -g\cdot Y(t) \\ \dot{Var}(t) = -2g\cdot Var(t) + (2\gamma - g)Y(t) \tag{8.12}$$

under suitable initial conditions $Y(t=0) = 1$, $Var(t=0) = 0$. The solutions are

$$Y(t) = e^{-g(t-t_0)} \\ Var(t) = \frac{(2\gamma-g)}{g}e^{-g(t-t_0)}\left(1 - e^{-g(t-t_0)}\right) \quad . \tag{8.13}$$

These equations will be investigated further in Sec. 8.3, looking at a continuum of pathogenicities, not just two-strain dynamics of one strain with finite pathogenicity versus one with zero pathogenicity.

In a simplified model, where the SIR subsystem is assumed to be stationary (due to its fast dynamics), we can show analytically the divergence of variance and power law behavior for the size of the epidemics $p(X)$ as soon as the pathogenicity is going towards zero. Hence the counter intuitively large number of disease cases in some realizations of the process can be understood as large scale fluctuations in a critical system with order parameter ε towards zero.

The master equation for YX in stationary SIR results in a birth–death process

$$\begin{aligned}\frac{d}{dt}p(Y,X,t) &= (b\cdot(Y-1)+c)\,p(Y-1,X,t) \\ &\quad + a\cdot(Y+1)\,p(Y+1,X,t) + g\cdot Y\,p(Y,X-1,t) \\ &\quad - (bY+aY+gY+c)\,p(Y,X,t) \quad . \end{aligned} \tag{8.14}$$

For the size distribution of the epidemic we obtain power law behavior

$$p_\varepsilon(X) := \lim_{t\to\infty} p(Y=0,X,t) \sim \frac{1}{2\sqrt{\pi\beta}}\cdot\varepsilon^{\frac{1}{2}}\cdot X^{-\frac{3}{2}} \quad . \tag{8.15}$$

for $\varepsilon \longrightarrow 0$ and large X. This was obtained by approximations to a solution with the hypergeometric function

$$p_\varepsilon(X) = \sqrt{\varepsilon}\cdot\frac{2^{-(X+1)}}{\sqrt{\beta}}\cdot {}_2F_1\left(\frac{3-X}{2},\frac{2-X}{2};2;1-\frac{\varepsilon}{\beta}\right). \tag{8.16}$$

For the detailed calculations see Appendix F. This behavior near criticality is also observed in the full SIRYX-system in simulations where the pathogenicity ε is small, i.e. in the range of the mutation rate μ.

In spatial versions of this model it is expected that the critical exponents are those of directed percolation (private communication, H.K. Jansen, Düsseldorf, see also [Janssen (1981)]).

8.3 Evolution toward Criticality

Furthermore, the epidemiological system with mutants of accidental pathogens is driven by evolution toward the critical threshold of small pathogenicity and hence to large critical fluctuations [Stollenwerk and Jansen (2003,b)]. The mechanism is simply the disadvantage of the more harmful strains against their less harmful opponents as they remove their hosts from the system and prevent themselves from spreading the respective harmful mutants further. Only mild mistakes can survive for a long period of time, possibly very long for very mildly pathogenic strains. Thus you arrive at the seemingly paradoxical situation that by reducing the pathogenicity by a factor of ten you actually often observe higher numbers of disease cases X (see Fig. 8.1, a)). The paradox is of course resolved by inspecting the number of mutant infected hosts Y which increases greatly by reducing the pathogenicity (see Fig. 8.1b). This qualitative explanation of why the mildly harmful mutants are dominating the epidemiology of accidental pathogenes has been proved quantitatively in [Stollenwerk and Jansen (2003,b)].

Hence, comparing the simple SIRX model with occasional transitions into the disease class and the multi-mutant model of SIRYX-type, we find for the simpler SIRX model a continuous increase in cumulative disease cases (see Fig. 8.2, a)), whereas in the refined SIRYX-model we find large outbreaks of disease and rather silent phases in between. This gives the step-like curve in cumulative disease cases for the SIRYX model (see Fig. 8.2b)).

How this difference between the different models translates under realistic circumstances of large populations modeling notification data for meningococcal disease in different countries, we will investigate in the next section.

8.3.1 *Fast and slow time scales in the meningitis model*

In sandpile models two time scales appear: the slow time scale of sand toppling onto the sandpile and the fast time scale of avalanches running down the pile. In simulations and theoretical models often these time scales are separated by making the avalanches infinitely fast. Likewise, in our epidemic model we have the slow time scale of mutations of different bacteria strains characterized by the mutation rate μ, and the fast time scale of the infection process for these strains characterized by the contact rate β (and rates α and γ). The former parameter ε, the pathogenicity, will now become a state variable which the system adjusts on its own, like the slope of the sandpile being adjusted on its own in the paradigmatic model of SOC.

We analyze the situation of a finite mutation rate and find an analytic solution in terms of an infinite sum. This solution shows evolution towards criticality. The above simplifications of stationarity of the SIR subsystem

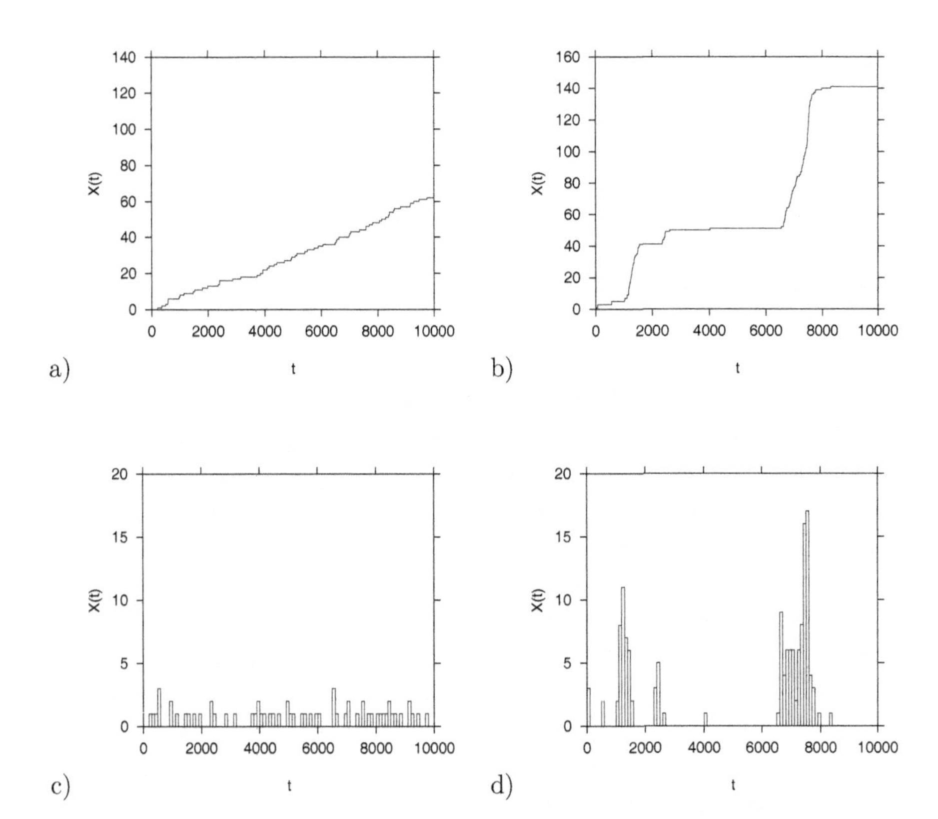

Fig. 8.2 A comparison between the simple SIRX model and the multi-mutant SIRYX model shows low but continuous incidence numbers for the SIRX system in c) whereas outbreaks are visible for the SIRYX system in d). The cumulative number of cases curves are shown in a) and b) as described in the text.

are used to obtain analytic solutions. Simulations of the full SIRYX subsystem agree well with the analytic results [Stollenwerk and Jansen (2003,a)].

8.3.2 *Analytics for evolution toward criticality*

We assume that no mutant infected hosts are present initially, but with mutation rate μ mutants with equally distributed pathogenicity are created. We take Eq. (8.9) and integrate with the result of

$$\langle Y\rangle(\varepsilon, t) = \frac{c}{g}\left(1 - e^{-gt}\right)$$

with import rate $c = \mu I^* \cdot S^*/N = \mu I^* \frac{\gamma}{\beta}$, hence $c/g = \mu I^*/\varepsilon$. This is derived from the ODE $\frac{d}{dt}\langle Y\rangle = (b-a)\langle Y\rangle + c$. The solution can now only be obtained numerically (or with $\int (e^x/x)\,dx$ as given function).

The probability to find ε at time t is $p(\varepsilon, t)$, using Eq. (8.17),

$$p(\varepsilon,t) := \frac{\langle Y\rangle(\varepsilon,t)}{\int_0^{\varepsilon_m} \langle Y\rangle(\varepsilon,t)\,d\varepsilon} = \frac{\frac{1}{\varepsilon}\left(1-e^{-\varepsilon\frac{\gamma}{\beta}t}\right)}{\int_0^{\varepsilon_m} \frac{1}{\varepsilon}\left(1-e^{-\varepsilon\frac{\gamma}{\beta}t}\right)\,d\varepsilon} \quad . \tag{8.17}$$

With the substitution $z := -\varepsilon\frac{\gamma}{\beta}t$, hence $\frac{dz}{d\varepsilon} = -\frac{\gamma}{\beta}t$ it is

$$\int_0^{\varepsilon_m} \frac{1}{\varepsilon}\left(1-e^{-\varepsilon\frac{\gamma}{\beta}t}\right)\,d\varepsilon = \lim_{\varepsilon_0\to 0}\left(\int_{\varepsilon_0}^{\varepsilon_m}\frac{1}{\varepsilon}\,d\varepsilon - \int_{\varepsilon_0}^{\varepsilon_m}\frac{1}{\varepsilon}e^{-\varepsilon\frac{\gamma}{\beta}t}\,d\varepsilon\right) \quad . \tag{8.18}$$

The integration can be performed by using the exponential integral function $\mathrm{Ei}(y) := \int_{-\infty}^{y} \frac{e^z}{z}\,dz$ and its representation by an infinite sum $\mathrm{Ei}(y) = \ln|y| + \sum_{\nu=1}^{\infty}\frac{y^\nu}{\nu\cdot\nu!} + C$. It is explicitly

$$\int_0^{\varepsilon_m} \frac{1}{\varepsilon}\left(1-e^{-\varepsilon\frac{\gamma}{\beta}t}\right)\,d\varepsilon = \lim_{\varepsilon_0\to 0}\left(\ln(\varepsilon_m) - \ln(\varepsilon_0) + \int_{-\frac{\gamma}{\beta}t\varepsilon_0}^{-\frac{\gamma}{\beta}t\varepsilon_m}\frac{1}{z}e^z\,dz\right) \tag{8.19}$$

and taking the limit

$$\lim_{\varepsilon_0\to 0}\left(\ln(\varepsilon_m) - \ln(\varepsilon_0) + \mathrm{Ei}\left(-\frac{\gamma}{\beta}t\varepsilon_0\right) - \mathrm{Ei}\left(-\frac{\gamma}{\beta}t\varepsilon_m\right)\right) \tag{8.20}$$

we obtain the finite sum representation

$$\int_0^{\varepsilon_m} \frac{1}{\varepsilon}\left(1-e^{-\varepsilon\frac{\gamma}{\beta}t}\right)\,d\varepsilon = \sum_{\nu=1}^{\infty}(-1)^{\nu+1}\frac{\left(\varepsilon_m\frac{\gamma}{\beta}t\right)^\nu}{\nu\cdot\nu!} \quad . \tag{8.21}$$

The result of the distribution of the pathogenicity is finally given by

$$p(\varepsilon,t) = \frac{\frac{1}{\varepsilon}\left(1-e^{-\varepsilon\frac{\gamma}{\beta}t}\right)}{\sum_{\nu=1}^{\infty}(-1)^{\nu+1}\cdot\frac{\left(\varepsilon_m\frac{\gamma}{\beta}t\right)^\nu}{\nu\cdot\nu!}} \quad . \tag{8.22}$$

The distribution $p(\varepsilon, t)$ from Eq. (8.25) is shown for three different times in Fig. 8.3.

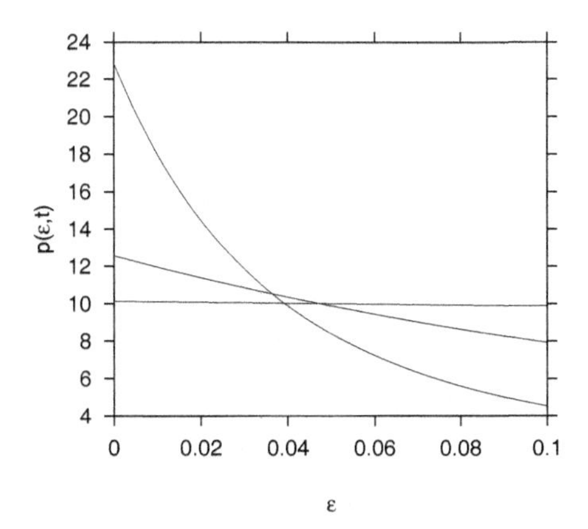

Fig. 8.3 Theoretical curves for times $t = 1$, $t = 20$, $t = 100$ as in Eq (8.25). For increasing time the distribution $p(\varepsilon, t)$ goes up for small pathogenicities ε and decreases for large ε.

8.3.3 *Simulation for evolution toward criticality*

In simulations for our model we start with a resident strain with vanishing pathogenicity and allow for mutations to various strains with different pathogenicities ε_i. For all strains the mutation rate is the same. We consider the distribution of the various ε_i in a population of mutant infected over time (see Eq. (8.23)) and compare with theoretical results in Fig. 8.4. The distribution of pathogenicities in an ensemble of hosts infected with different mutant strains is given by

$$\hat{p}(\varepsilon_i, t) := \frac{\sum_j Y_j(\varepsilon_i, t)}{\sum_i \sum_j Y_j(\varepsilon_i, t) \cdot \Delta\varepsilon} \tag{8.23}$$

with $\Delta\varepsilon$ the length of the considered ε-interval times the number of ε-values.

Remarkably, the result for the lowest value of $\varepsilon = 0.002$ still is in very good agreement with the theoretical curve even though ε is only twenty times larger than the mutation rate μ [Stollenwerk and Jansen (2003,a)].

8.3.4 *Spatial modeling: Outbreak clusters without direct disease contacts*

The SIRYX model is simulated on a two-dimensional grid with periodic boundary conditions. 32 $\times$ 32 individuals have been simulated.

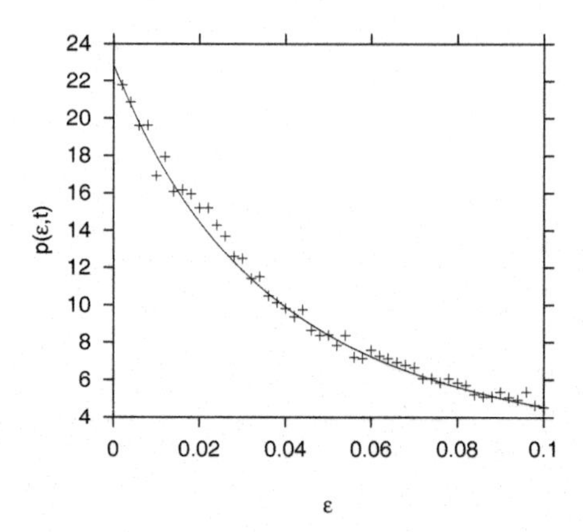

Fig. 8.4 Comparison between model and simulations for time $t = 100$. Each data point of the simulation is an average over 5000 runs, and 50 values for ε are taken.

The simulation we present here (Fig. 8.5) gives an example for two disease cases being linked by a sea of mutant infected, but not by direct contact. In the realization we show here, the epidemic with mutant infected has already been going on for quite a while; the two cases appear rather close in time but are separated spatially by several intermediate individuals.

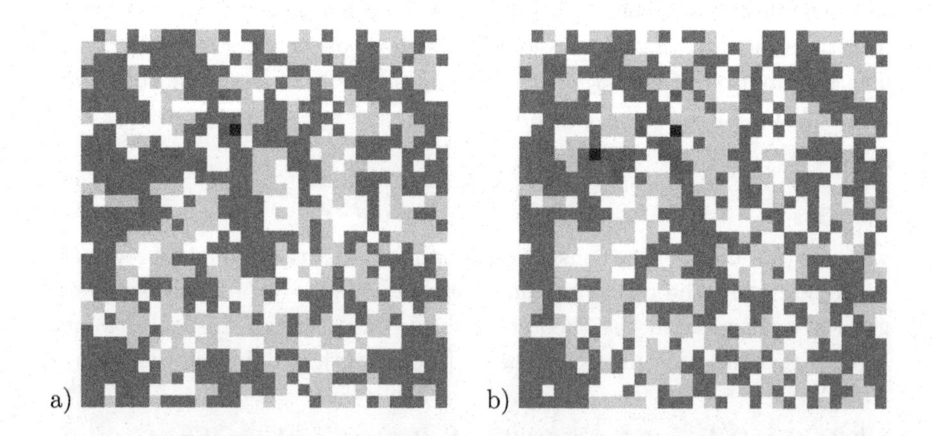

Fig. 8.5 Two pictures from a spatial simulation of the SIRYX model, showing the system when a) the first disease case and b) the second disease case appear. Shades indicate the different host classes. Of importance are dark for mutant-infected and black for disease cases.

In another spatial simulation (Fig. 8.6) we vary the pathogenicity of the mutant infected between 0 and $\varepsilon_{max} := 0.1$, shading in dark between grey for low and black for high pathogenicity. We also show the time series of the mean pathogenicity of the mutant infected. Although we find huge fluctuations, the general tendency of decreasing pathogenicity is clearly visible.

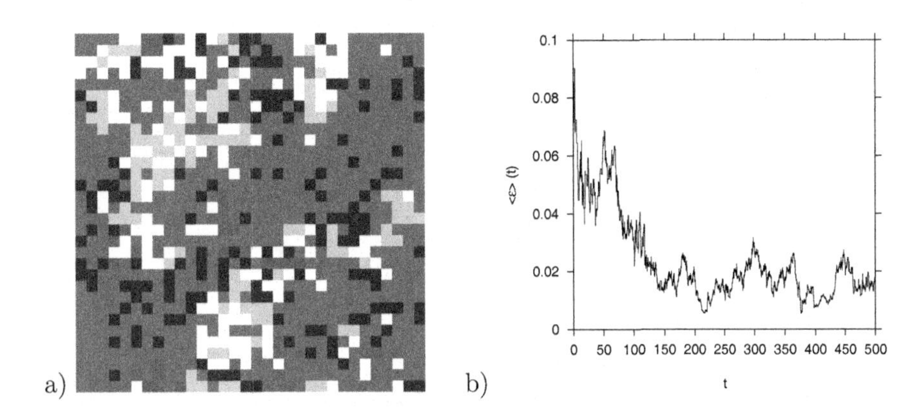

Fig. 8.6 a) Randomly varying pathogenicity, shaded in dark grey to black for decreasing pathogenicity. b) Mean value of pathogenicity over time.

8.3.5 *Mean pathogenicity goes toward zero*

The probability of finding ε at time t is $p(\varepsilon, t)$, as seen before,

$$p(\varepsilon, t) := \frac{\langle Y \rangle(\varepsilon, t)}{\int_0^{\varepsilon_m} \langle Y \rangle(\varepsilon, t)\, d\varepsilon} = \frac{\frac{1}{\varepsilon}\left(1 - e^{-\varepsilon \frac{\gamma}{\beta} t}\right)}{\int_0^{\varepsilon_m} \frac{1}{\varepsilon}\left(1 - e^{-\varepsilon \frac{\gamma}{\beta} t}\right) d\varepsilon} \tag{8.24}$$

with the result

$$p(\varepsilon, t) = \frac{\frac{1}{\varepsilon}\left(1 - e^{-\varepsilon \frac{\gamma}{\beta} t}\right)}{\sum_{\nu=1}^{\infty} (-1)^{\nu+1} \cdot \frac{\left(\varepsilon_m \frac{\gamma}{\beta} t\right)^{\nu}}{\nu \cdot \nu!}} \quad . \tag{8.25}$$

Thus the mean pathogenicity is defined as

$$\langle \varepsilon \rangle(t) := \int_0^{\varepsilon_m} \varepsilon\, p(\varepsilon, t)\, d\varepsilon \tag{8.26}$$

hence

$$\langle\varepsilon\rangle(t) = \frac{\varepsilon_m - \frac{1-e^{-\varepsilon_m \frac{\gamma}{\beta} t}}{\frac{\gamma}{\beta} t}}{\sum_{\nu=1}^{\infty}(-1)^{\nu+1} \cdot \frac{\left(\varepsilon_m \frac{\gamma}{\beta} t\right)^{\nu}}{\nu \cdot \nu!}} \tag{8.27}$$

It is expected that the mean pathogenicity goes towards zero, as the simulation Fig. 8.6b) suggests. However, the Eq. (8.27) is numerically difficult to evaluate in the present form.

Instead of using the integral representation of the exponential integral function as before

$$\int_0^{\varepsilon_m} \frac{1}{\varepsilon}\left(1 - e^{-\varepsilon \frac{\gamma}{\beta} t}\right) \, d\varepsilon = \sum_{\nu=1}^{\infty}(-1)^{\nu+1} \cdot \frac{\left(\varepsilon_m \frac{\gamma}{\beta} t\right)^{\nu}}{\nu \cdot \nu!} \tag{8.28}$$

one can approximate for large t

$$\int_0^{\varepsilon_m} \frac{1}{\varepsilon}\left(1 - e^{-\varepsilon \frac{\gamma}{\beta} t}\right) \, d\varepsilon \approx 1 + \ln\left(\varepsilon_m \frac{\gamma}{\beta} t\right) \quad . \tag{8.29}$$

Hence the mean pathogenicity being defined as $\langle\varepsilon\rangle(t) := \int_0^{\varepsilon_m} \varepsilon \; p(\varepsilon, t) \; d\varepsilon$ becomes

$$\langle\varepsilon\rangle(t) \frac{\varepsilon_m - \frac{1-e^{-\varepsilon_m \frac{\gamma}{\beta} t}}{\frac{\gamma}{\beta} t}}{\sum_{\nu=1}^{\infty}(-1)^{\nu+1} \cdot \frac{\left(\varepsilon_m \frac{\gamma}{\beta} t\right)^{\nu}}{\nu \cdot \nu!}} \approx \frac{\varepsilon_m - \frac{1-e^{-\varepsilon_m \frac{\gamma}{\beta} t}}{\frac{\gamma}{\beta} t}}{1 + \ln\left(\varepsilon_m \frac{\gamma}{\beta} t\right)} \tag{8.30}$$

which can be evaluated numerically.

The approximation

$$\int_0^{\varepsilon_m} \frac{1}{\varepsilon}\left(1 - e^{-\varepsilon \frac{\gamma}{\beta} t}\right) \, d\varepsilon \approx 1 + \ln\left(\varepsilon_m \frac{\gamma}{\beta} t\right) \tag{8.31}$$

is derived as follows.

Cut the interval of the integration in two pieces, defining $\check{\varepsilon} := \frac{\beta}{\gamma t}$, and use different approximations for $e^{-\varepsilon \frac{\gamma}{\beta} t}$ in each of the two integrals. This gives

$$\int_0^{\varepsilon_m} \frac{1}{\varepsilon}\left(1 - e^{-\varepsilon \frac{\gamma}{\beta} t}\right) \, d\varepsilon = \int_0^{\check{\varepsilon}} \frac{1}{\varepsilon}\left(1 - e^{-\varepsilon \frac{\gamma}{\beta} t}\right) \, d\varepsilon + \int_{\check{\varepsilon}}^{\varepsilon_m} \frac{1}{\varepsilon}\left(1 - e^{-\varepsilon \frac{\gamma}{\beta} t}\right) \, d\varepsilon$$

$$
\begin{aligned}
&= \int_0^{\check{\varepsilon}} \frac{1}{\varepsilon}\left(1 - \sum_{\nu=1}^{\infty} \frac{\left(-\varepsilon\frac{\gamma}{\beta}t\right)^{\nu}}{\nu!}\right) d\varepsilon + \int_{\check{\varepsilon}}^{\varepsilon_m} \frac{1}{\varepsilon}\left(1 - \frac{1}{\sum_{\nu=1}^{\infty} \frac{\left(\varepsilon\frac{\gamma}{\beta}t\right)^{\nu}}{\nu!}}\right) d\varepsilon \qquad (8.32)\\
&\approx \int_0^{\check{\varepsilon}} \frac{1}{\varepsilon}\left(1 - \left(1 - \varepsilon\frac{\gamma}{\beta}t\right)\right) d\varepsilon + \int_{\check{\varepsilon}}^{\varepsilon_m} \frac{1}{\varepsilon}\left(1 - \frac{1}{1+\varepsilon\frac{\gamma}{\beta}t}\right) d\varepsilon\\
&= \int_0^{\check{\varepsilon}} \frac{\gamma}{\beta}t \, d\varepsilon + \int_{\check{\varepsilon}}^{\varepsilon_m} \frac{1}{\varepsilon}\, d\varepsilon\\
&= \frac{\gamma}{\beta}t \cdot \check{\varepsilon} + \ln\left(\frac{\varepsilon_m}{\check{\varepsilon}}\right)\\
&= 1 + \ln\left(\varepsilon_m \frac{\gamma}{\beta}t\right) \qquad ,
\end{aligned}
$$

using again $\check{\varepsilon} := \frac{\beta}{\gamma t}$ in the last step. This proves the approximation for large t.

Using the approximation for large t and using the infinite sum representation for small t one can finally plot the mean pathogenicity curve for a large time scale (see Fig. 8.7).

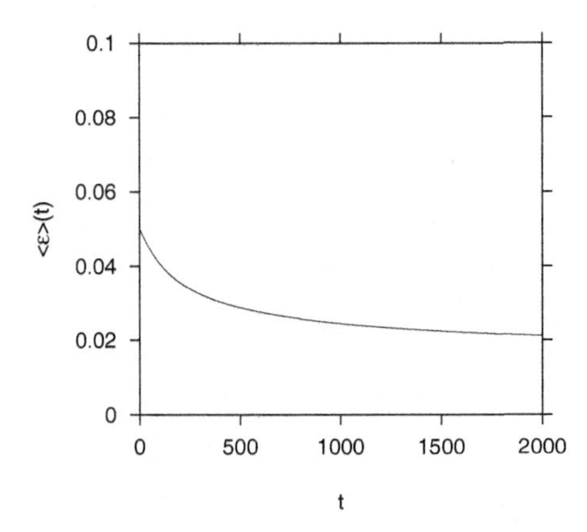

Fig. 8.7 Mean of pathogenicity decreases. The curve for larger times uses Paco'sthe approximation with additional power series term: $1 - 0.4 + \ln\left(\varepsilon_m \frac{\gamma}{\beta}t\right)$ which is approximately $1 - 0.25 + \ln\left(\varepsilon_m \frac{\gamma}{\beta}t\right)$. Cutting point here is at $t = 800$.

This shows that the mean pathogenicity decreases very slowly, as was also previously suggested by the stochastic simulation of the system.

8.4 Empirical Data Show Fast Epidemic Response and Long-Lasting Fluctuations

The first information to include in a model on menigococcal disease is the observation from studies not only of disease cases X but also studies of harmless carriage I in school classes [Cartwright (1995)], namely that natural carriage is between 10 and 40%. This sets the ratio between basic epidemiological parameters, as described in detail in [Stollenwerk and Jansen (2003,a)], but leaves complete freedom to the overall time scale. Hence time in Figs. 8.1 and 8.2 is given in arbitrary units of simulation time.

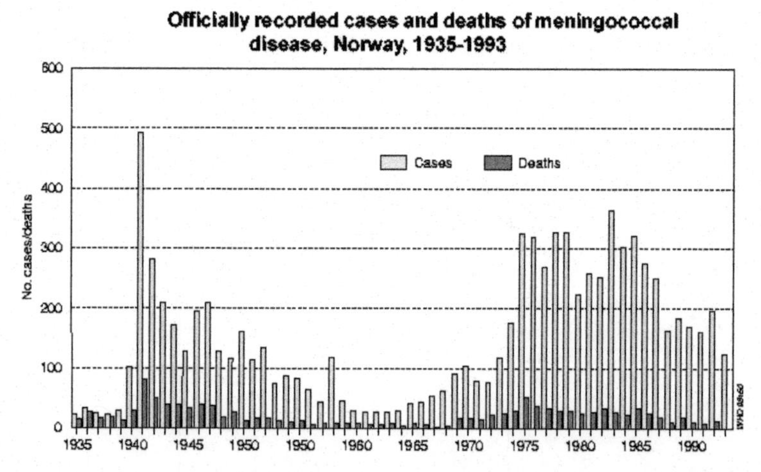

Fig. 8.8 Yearly cases of menigoccocal disease for Norway, notification data (World Health Organization (WHO), http://www.who.int/emc, document WHO/EMC/BAC/98.3). Decade-long outbreaks are visible.

A first inspection of empirical data on outbreak patterns of meningococcal disease is puzzling. On the one hand side, in long time series for a country like Norway one observes decade-long outbreaks (see Fig. 8.8), suggesting that basic epidemiological parameters like inverse infection and recovery rate, etc., are of the order of several months, possibly as long as one year.

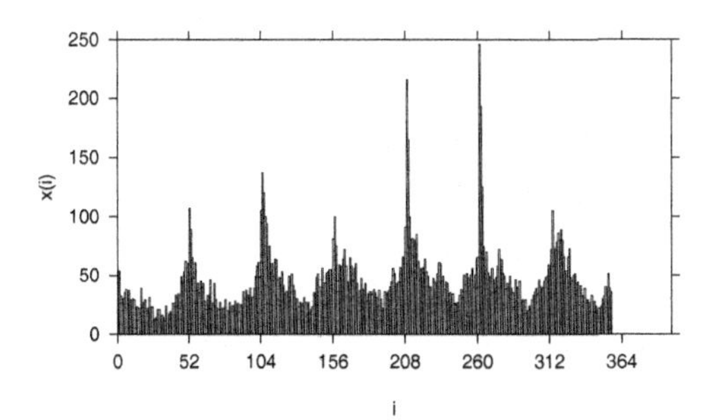

Fig. 8.9 Weekly data of notified cases of meningococcal disease in England and Wales. A strong seasonality is visible. Time is given in weeks, starting 1 January, 1995.

On the other hand, in recently obtained single case data from England and Wales, binned into weekly data, a strong seasonal pattern in meningococcal disease notifications is clearly visible, with superimposed very strong outbreaks around Christmas and the change of year (see Fig. 8.9). A similar pattern is visible for the nine regions in which England and Wales are divided under public health notification, an always strong seasonality is often accompanied by high Christmas peaks, the regions being of similar population size as Norway with around 5 *million* inhabitants. Assuming a seasonal forcing of the contact rate coinciding with the seasonality in climate and hence meeting pattern in the underlying population, this leaves only a time scale of quick adjustment of the infection process for parameters like inverse infection and recovery rate, etc., in the range of a few weeks. On top of the Christmas peak, a strong increase of cases in the 52nd week of the calendar year and higher incidence rates also in the two following weeks, the first and second week in January, even suggests a shorter time scale of two to four weeks.

A possible explanation for this apparent contradiction could simply be different strains acting on different time scales, and even in different countries. Although this hypothesis has been expressed, the vast amount of mutants found and scarce microbiological data available to classify them, resulting in mechanistic models of transition to disease. These could not isolate single strains responsible for harmless infection, those causing frequent disease cases, and those causing fluctuations of disease cases on longer time scales. We cannot rule out such mechanisms but, however, we were taken by surprise, that a rather restricted model such as the SIRYX model described above, can capture both the quick response to seasonal forcing, as will be shown below, but also due to its closeness to a critical threshold can produce huge long term fluctuations on the time scale of decades when compared to the given time scale of a year given by seasonality. In contrast, the simpler SIRX model, being forced seasonally, still only can give rise to fluctuations predicted by a Poisson process; technically giving a variance only being in the range of the mean, but not showing the much larger and time-correlated critical fluctuations of the SIRYX model.

Interestingly, whereas any distinction between the SIRX and the SIRYX model would be very difficult on the basis of the short term weekly data from England and Wales, the distinction is quite easy for long term simulations exploiting the critical fluctuations.

8.4.1 *Modeling fast epidemic response finds long-lasting fluctuations*

To model the seasonal data from England and Wales, we first observe data from the nine regions, into which England and Wales is divided. By taking the mean, weighted with the total number of cases in each region over the observation period, we can reduce the effect of the pronounced Christmas peak, which we exclude from further modeling now.

Secondly, we adjust the parameters of the simple SIRX model, respectively the multi-mutant SIRYX model, to the seasonality and the noise level of the weighted mean data set. Starting from the stationary state solution for the SIRYX model with constant time independent contact rate we obtained good visual agreement between model and data using a parameter set with fixed ratio of susceptible, infected and recovered. Hence we fix the ratio of the basic epidemic parameters α, β and γ of the SIR subsystem, and we fix the mutation rate μ and the pathogenicity ε to roughly obtain the noise level of the observed data. Finally, we fixed the absolute value of γ to the time scale given by the data's seasonality, especially the slight shift which means fast response, to exact solar seasonal forcing in the contact rate. This left us with an upper limit of inverse recovery $\gamma^{-1} = 4$ *weeks*, giving a minimum of disease cases X about seven weeks after midsummer, as observed in the data. Temperature insecurities and related mixing paterns of the host population could change this picture in the range of plus or minus two weeks, but not allowing a response in the range of months or years, which would result in the smoothing out the of the seasonality.

The ODE system for the SIRYX model with seasonality and two harmful competing mutants, with ε_0 as resident strain and ε for the more harmful

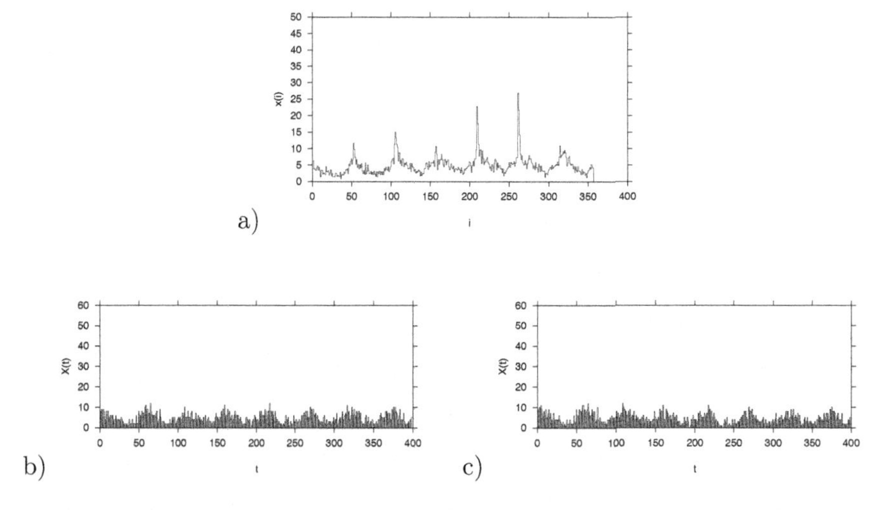

Fig. 8.10 Comparison between a) data from England and Wales and simulations with b) the SIRX model and c) the SIRYX model. In a) the weighted mean over nine regions of England and Wales is shown for the seven years of weekly data. b) shows a simulation of the simple SIRX, parameters adjusted to qualitatively match the data in Fig. 8.10a), for a comparable amount of time. c) shows a simulation of the multi-mutant SIRYX-model, taking the same basic parameters of the SIRX model and further adjustments of the additional parameters into account to match the data. Population size is $N = 5$ million, roughly the size of a typical region in England and Wales. Both models resemble the data fairly well in its seasonality and noise level, not attempting to also model the Christmas peak. Little difference is visible between the models.

mutant, reads as

$$\begin{aligned}
\frac{dS}{dt} &= \alpha \cdot R - (\beta(t) - \varepsilon_0) \cdot \frac{S}{N} I - (\beta(t) - \varepsilon) \cdot \frac{S}{N} Y + \varphi X \\
\frac{dI}{dt} &= (\beta(t) - \mu - \varepsilon_0) \cdot \frac{S}{N} I - \gamma I + \nu \frac{S}{N} Y \\
\frac{dR}{dt} &= \gamma \cdot (I + Y) - \alpha R \\
\frac{dY}{dt} &= (\beta(t) - \nu - \varepsilon) \cdot \frac{S}{N} Y - \gamma Y + \mu \cdot \frac{S}{N} I \\
\frac{dX}{dt} &= \varepsilon \cdot \frac{S}{N} Y + \varepsilon_0 \cdot \frac{S}{N} I - \varphi X
\end{aligned} \tag{8.33}$$

with seasonally-forced infection rate $\beta(t) = \beta_0 \left(1 + \beta_1 \cos\left(\frac{2\pi}{T} t\right)\right)$ and with $T = 52$ *weeks*. Furthermore, we set $\nu := \mu$, $\varepsilon := \mu$, which means that the resident strain has pathogenicity in the range of the mutation rate, and $\varphi := 0$.

For the stochastic SIRYX system the transition probabilities $w_{\underline{\tilde{n}},\underline{n}}$ are then given (omitting unchanged indices in $\underline{\tilde{n}}$, with respect to $\underline{n}$) by

$$\begin{aligned}
w_{(R-1,S+1),(R,S)} &= \alpha \cdot R &,&\quad R \xrightarrow{\alpha} S \\
w_{(S-1,I+1),(S,I)} &= (\beta(t) - \mu - \varepsilon_0) \cdot \tfrac{I}{N}\, S &,&\quad S + I \xrightarrow{\beta-\mu} I + I \\
w_{(S-1,Y+1),(S,Y)} &= \mu \cdot \tfrac{I}{N}\, S &,&\quad \xrightarrow{\mu} Y + I \\
w_{(I-1,R+1),(I,R)} &= \gamma \cdot I &,&\quad I \xrightarrow{\gamma} R \\
w_{(S-1,Y+1),(S,Y)} &= (\beta(t) - \nu - \varepsilon) \cdot \tfrac{Y}{N}\, S &,&\quad S + Y \xrightarrow{\beta-\nu-\varepsilon} Y + Y \\
w_{(S-1,I+1),(S,I)} &= \nu \cdot \tfrac{Y}{N}\, S &,&\quad \xrightarrow{\nu} I + Y \\
w_{(S-1,X+1),(S,X)} &= \varepsilon \cdot \tfrac{Y}{N}\, S &,&\quad \xrightarrow{\varepsilon} X + Y \\
w_{(Y-1,R+1),(Y,R)} &= \gamma \cdot Y &,&\quad Y \xrightarrow{\gamma} R \\
w_{(X-1,S+1),(X,S)} &= \varphi \cdot X &,&\quad X \xrightarrow{\varphi} S
\end{aligned} \tag{8.34}$$

along with the respective reaction schemes.

Parameters are $\beta = 0.3125$, $\alpha = 0.25$, $\gamma = 0.25$, $\beta_1 = 0.3$, $\mu = 0.000006$, $\nu = 0.000006$, $\varepsilon = 0.005$, $\phi = 300$, $t_{max} = 400$, $T = 52$, $\varepsilon_0 = 0.000006$.

The SIRX model uses the same basic epidemic parameters α, β and γ as the SIRYX model. No mutation rate is needed here, since we only have one strain of pathogenes in this model, and an adjusted pathogenicity accounts for the lack of mutants Y in this model. As shown in Fig. 8.10 there are hardly any differences visible between the SIRX model and the SIRYX model on this time scale, both of which describe the mean regional data in England and Wales quite well in terms of seasonality and noise level. So the question arises, if the differences between the SIRX model and the SIRYX model, which are clearly present as shown in Fig. 8.2, could be detected under other circumstances, could we eventually decide between the two models by observational data? The first candidate could be the population size: a larger population size might show a qualitative difference between the two models due to averaging out the visually disturbing flucutations because of small numbers of disease cases. Hence we can try a ten times larger population of $N = 50$ *million*, which roughly corresponds to the total population size of England and Wales, comparable with the data shown in Fig. 8.9. The results of our simulations are shown in Fig. 8.11. But again there is hardly any difference visible between the two models. The only slight hint of larger fluctuations could be the steeper rise of disease cases in the SIRYX model, Fig. 8.11b), around week 300 and 320, as compared to the SIRX model, Fig. 8.11a). However, it would be hard to test on such subtle differences under the present noise level of the data, Fig. 8.9, rigorously.

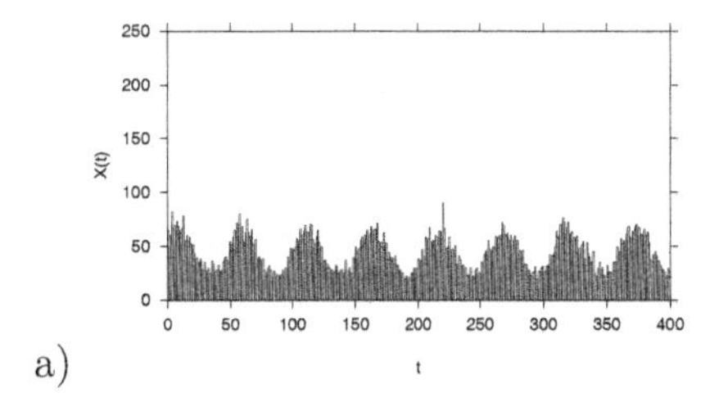

a)

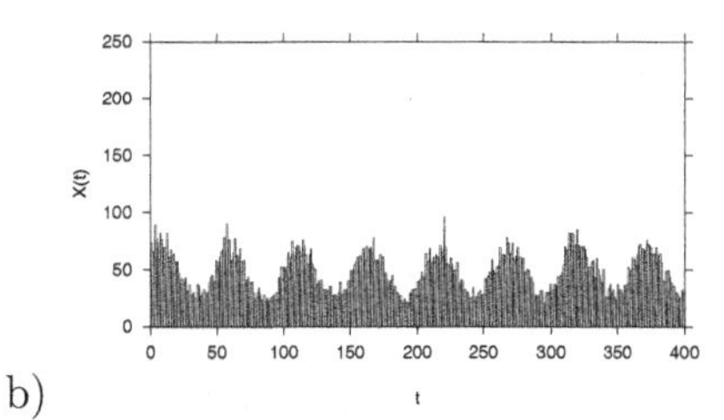

b)

Fig. 8.11 Simulation of a) the SIRX model and b) the SIRYX model with the same parameter sets as in Fig. 8.10 but larger host population size $N = 50\ 000\ 000$, hence approximately the size of England and Wales. Only minor differences are visible; too weak to be tested on with empirical data.

Hence we have to look at a different time scale in order to see any profound difference between the models, not just a different scale in system size. Therefore, we performed a comparative study, binning the number of disease cases not into weeks but years (keeping the weekly time scale in the axis description in order to remind us of the parameters adjusted to the weekly data, and being able to compare the longer time duration of the simulations), increasing the simulated time to roughly 1200 weeks (corresponding to 23 years), three times longer than the previous simulations and also than the empirical data.

The result is shown in Fig. 8.12. In Fig. 8.12a) the SIRX model for 5 million population size shows quite some fluctuations from year to year, whereas the SIRYX model in Fig. 8.12b) for the same system size sometimes shows much larger variabilty, but sometimes not. For example, between weeks 400 and 800 it would be quite difficult to distinguish the two realizations shown here. For population size ten times larger, corresponding to the size of England and Wales, the differences between SIRX model in 8.12c) and SIRYX model in 8.12d) is even less pronounced over the entire simulation time. So again any testing between the models would face severe difficulties, the more since our data sets from England and Wales are much shorter than the simulation times used here for the models. Hence only longer term data could help in this situation.

On the other hand, this set of simulations gives us a crucial hint from the theory of critical phenomena of how to proceed further in our analysis in so far as comparing the Figs. 8.12b) and d), the close to critical SIRYX model shows some autocorrelation in time in its fluctuations which also increases in length with system size. This is predicted by the theory of critical phenomena [Stanley (1971); Landau and Binder (2000)]). Namely, at criticality the autocorrelation time diverges and close to critcality the autocorrelation time increases as a power law. Renormalization theory should guarantee that pictures of the system look similar when changing system size and running time accordingly. This is the so called scaling of system size and time.

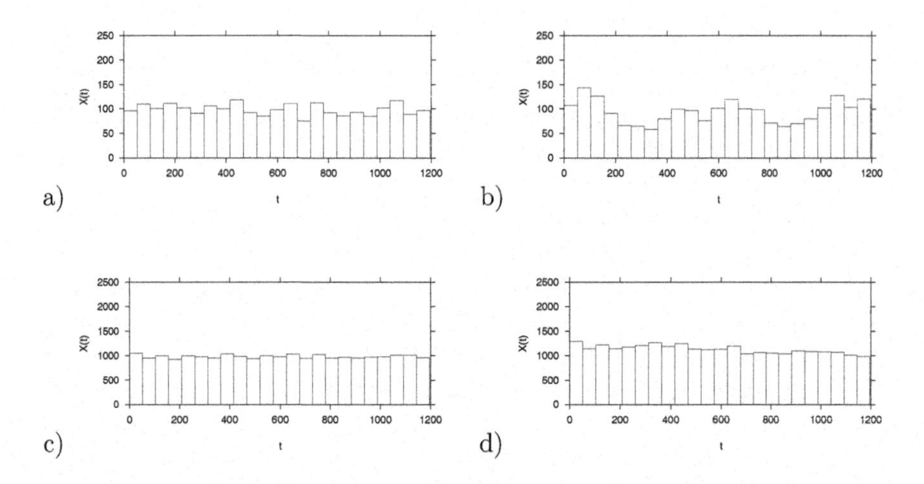

Fig. 8.12 Simulation of weekly cases, now binned into years for a) the SIRX model and b) the SIRYX model, for population size 5 *million*, c) and d) simulations of the above mentioned models now for population size 50 *million*. A detailed description of the graphs is given in the text.

Under the circumstances of Fig. 8.12, again a rigorous test would be difficult, since some autocorrelation in realizations of completely uncorrelated fluctuations is often obtained due to short time series. For example in the simulation of the SIRX system in Fig. 8.12a) one easily finds three subsequent years showing decreasing numbers of diseased cases. This occurs similarly in Fig. 8.12c). However, the autocorrelation functions for the data of Fig. 8.12 point into the same direction. Hence we perform even longer time simulations, expecting more pronounced fluctuations as time passes. Since our simulations are quite time consuming for large system sizes already at short time simulations, taking a day for most of the shown pictures so far, we perform longer simulations with just 1 *million* population size.

The results for four times longer simulations as in the previous Fig. 8.12 are shown in Fig. 8.13, comparing the SIRX model in a) and the SIRYX model in b). Though again for short periods, such as between week 1500 and 2000, there would be little difference between the models, the overall picture distinguishes between the models very well. Whereas the SIRX model in 8.13a) just shows minor fluctuations over the whole period of simulation, comparable essentially to a Poisson process, the SIRYX-model in 8.13b) shows large fluctuations and very surprisingly a huge epidemic between weeks 3000 and 3500 lasting around 12 years. This purely stochastic event could in real life easily be mistaken for an exogenously forced event, hence mathematically a drastic change in parameters, which it is clearly not in our case.

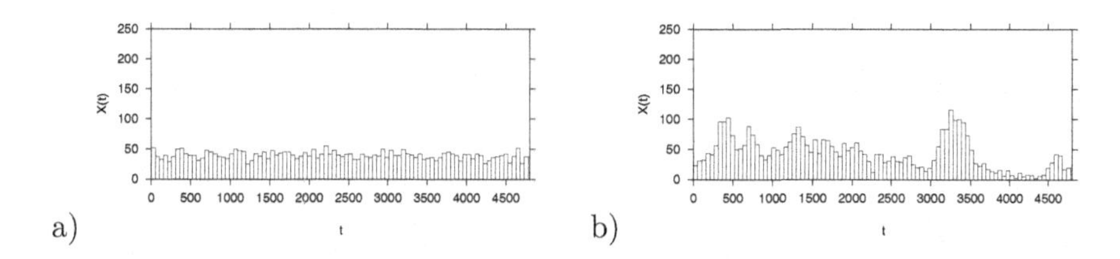

Fig. 8.13 Smaller population size $N = 1\ 000\ 000$ and $4\times$ longer time series. Nearly Poisson variance over mean for SIRX in a), it is 0.95. For SIRYX in b) it is 16.72, hence deviating significantly from the Poisson variance over mean ratio of 1.

To clear the point of stochastic decade-long outbreaks in our model of multi-mutant type further, we performed an even longer simulation of the SIRYX model (Fig. 8.14), finding over ten years very low levels of disease and thirty to fourty years of increased disease levels, a picture clearly resembling the situation in Norway of decade-long outbreaks of disease (Fig. 8.8). This would not be so surprising if our SIRYX model had been fitted to the Norwegian data directly, eventually implying a characteristic time scale of the basic epidemic parameters, α, β and γ in the range of months or years.

Given that these decade-long fluctuations in our model are not induced by the seasonality in the contact rate, but the closeness to criticality, we checked by simulations without seasonality, keeping the parameters otherwise as before. This again showed the huge decade-long fluctuations in disease level.

These findings have wide implications for public health concerns. Critical fluctuations as observed here can lead to long outbreaks of disease without any causal change in external factors (i.e. absence of parameter

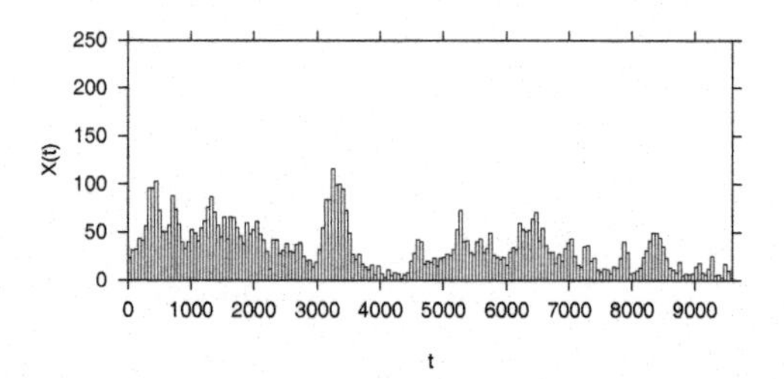

Fig. 8.14 Smaller population size $N = 1\ 000\ 000$ and $8\times$ longer time series for the SIRXY system. Now the variance over mean ratio is 15.94, again deviating significantly from a Poisson process. Large fluctuations take place also for longer times of simulation, hence are not only transient behavior.

change) and due only to fluctuations in hardly detectable mutant levels, an increase of Y cases.

8.4.2 *Data with fast epidemic response and long-lasting fluctuations*

Up to now we only have seen data from different countries, one of them showing fast epidemic response to seasonal forcing in contact rates and one other showing decade-long fluctuations. Recently available data from the USA, however, show on the one hand some seasonality (Fig. 8.15, monthly data for six years), which is not as clear as the British weekly data but still visible, and on the other hand huge decade-long fluctuations correlated over many years (Fig. 8.16, yearly data for 36 years), again not that pronounced as in the Norwegian data, but still clearly observable.

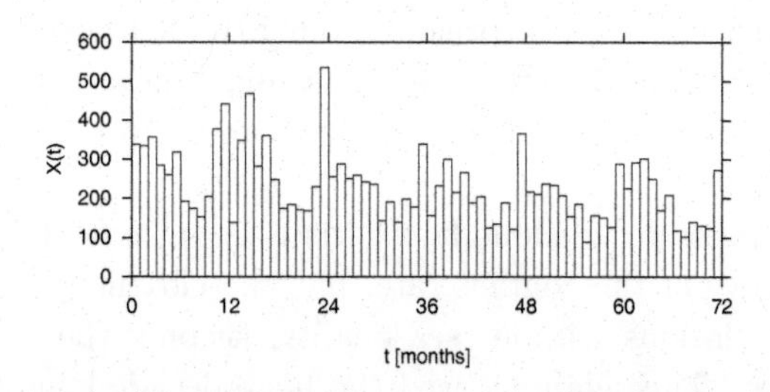

Fig. 8.15 Monthly data from the USA, 1996 to 2001, six years of data.

We have concentrated here on modeling the fast dynamics of meningitis data with strong seasonality, as visible in highly time resolved data from

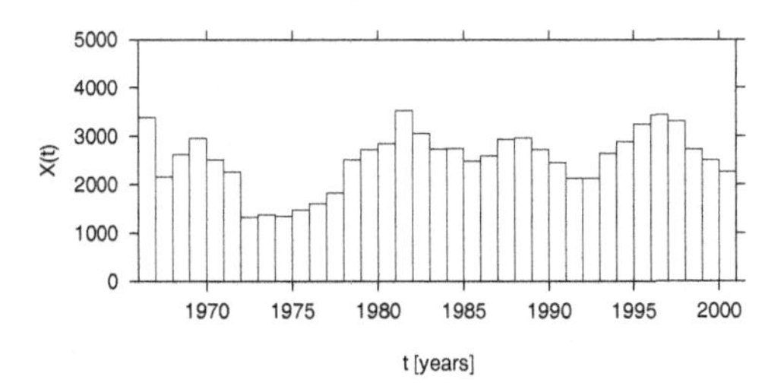

Fig. 8.16 Yearly data USA, 1966 to 2001, 36 years. yearly data have mean $\mu = 2519.7$ and standart deviation $\sigma = 581.9$, variance over mean $\frac{\sigma^2}{\mu} = 134.4$.

England and Wales, and found long term fluctuations, as seen in Norwegian long term data, without putting any new infromation into our model. In the case of data from the USA, both aspects are much weaker, hence not an ideal starting point for the analysis performed above, but still visible to a level that looks promising for future analysis.

Eventually, fine tuning of single parameters might be possible along the lines of earlier parameter estimation techniques with master equation simulations [Stollenwerk and Briggs (2000); Stollenwerk (2001)], for which the simulation time of the models has to be decreased significantly by approximations along the lines sketched in [Stollenwerk and Jansen (2003,a)], namely approximating the SIR part of the system deterministically.

In the previous chapters (Chapters 1 to 5) we indicated the techniques to analyze epidemic models towards criticality. These techniques give reliable and well-investigated results for very simple models like the SIS epidemic. Now that we have identified empirical epidemiological and evolutionary systems (Chapters 6 to 8) which indicate the relevance of critical fluctuations, the effort to rigorously apply the techniques beyond the mean field approximations from the previous chapters will be justified. Furthermore, the techniques from the earlier chapters will have much wider applications in quantitative biology.

Appendix A

Invariant Density of the Ulam Map

A.1 Time Evolution Equation of the Density

The Perron–Frobenius time evolution equation for stochastic systems

$$p_{n+1}(x_{n+1}) = \int_{-\infty}^{\infty}\int_{-\infty}^{\infty} \delta(x_{n+1} - f(x_n, \varepsilon_n))\, p_n(x_n) \cdot p_\varepsilon(\varepsilon_n)\, dx_n\, d\varepsilon_n \quad (A.1)$$

cannot only be used for noisy systems but also for the stochastic description of purely deterministic systems

$$x_{n+1} = f(x_n) \quad (A.2)$$

to calculate e.g. the invariant distribution p^*. With this result we can calculate even further quantities of the attractor, e.g. the Lyapunov exponent. Starting point is an ensemble $p_0(x_0)$ of starting values x_0 in the absence of noise.

For the Ulam map [Ulam and von Neumann (1947)]

$$x_{n+1} = 4x_n(1 - x_n) \quad (A.3)$$

the Perron–Frobenius equation on the unit interval is given by

$$p_{n+1}(x_{n+1}) = \int_0^1 \delta\left(x_{n+1} - 4x_n(1 - x_n)\right) \cdot p_n(x_n)\, dx_n \quad . \quad (A.4)$$

For the stationary distribution it is

$$p^*(x_{n+1}) = \int_0^1 \delta\left(x_{n+1} - 4x_n(1 - x_n)\right) \cdot p^*(x_n)\, dx_n \quad . \quad (A.5)$$

With this it can be easily verified that the distribution

$$p^*(x_n) = \frac{1}{\pi\sqrt{x_n(1-x_n)}} \tag{A.6}$$

satisfies the fixed point equation. The factor of π only is a result of the normalization of the distribution, i.e. $\int p^*(x_n)dx_n = 1$.

It is

$$\begin{aligned} p^*(x_{n+1}) &= \int_0^1 \delta\left(x_{n+1} - 4x_n(1-x_n)\right) \cdot \frac{1}{\pi\sqrt{x_n(1-x_n)}}\, dx_n \\ &= \int_{y(0)}^{y(\frac{1}{2})} \delta\left(x_{n+1} - 4y_n\right) \cdot \frac{1}{\pi\sqrt{y_n}} \cdot \frac{1}{1-2x_n}\, dy_n \\ &\quad + \int_{y(\frac{1}{2})}^{y(1)} \delta\left(x_{n+1} - 4y_n\right) \cdot \frac{1}{\pi\sqrt{y_n}} \cdot \frac{1}{1-2x_n}\, dy_n \end{aligned} \tag{A.7}$$

with the substitution $y_n := x_n(1-x_n) = x_n - x_n^2$; hence

$$dx_n = \frac{1}{\left(\frac{dy_n}{dx_n}\right)}\, dy_n = \frac{1}{1-2x_n}\, dy_n \quad .$$

Since the substitution is not invertible on the whole unit interval, meaning that

$$\begin{aligned} x &= \tfrac{1}{2} - \sqrt{\tfrac{1}{4} - y} \quad \text{for } x \in \left[0, \tfrac{1}{2}\right] \\ x &= \tfrac{1}{2} + \sqrt{\tfrac{1}{4} - y} \quad \text{for } x \in \left[\tfrac{1}{2}, 1\right] \end{aligned} \tag{A.8}$$

we have to split the integral into two branches and obtain

$$\begin{aligned} p^*(x_{n+1}) &= \int_{y(0)}^{y(\frac{1}{2})} \delta\left(x_{n+1} - 4y_n\right) \cdot \frac{1}{\pi\sqrt{y_n}} \cdot \frac{1}{1-2(\frac{1}{2}-\sqrt{\frac{1}{4}-y_n})}\, dy_n \\ &\quad + \int_{y(\frac{1}{2})}^{y(1)} \delta\left(x_{n+1} - 4y_n\right) \cdot \frac{1}{\pi\sqrt{y_n}} \cdot \frac{1}{1-2(\frac{1}{2}+\sqrt{\frac{1}{4}-y_n})}\, dy_n \quad . \end{aligned} \tag{A.9}$$

Hence it is

$$p^*(x_{n+1}) = \int_0^{\frac{1}{4}} \delta(x_{n+1} - 4y_n) \cdot \frac{1}{\pi\sqrt{y_n}} \cdot \frac{1}{\sqrt{1-4y_n}}\, dy_n$$

$$+ \int_{\frac{1}{4}}^{0} \delta(x_{n+1} - 4y_n) \cdot \frac{1}{\pi\sqrt{y_n}} \cdot \frac{1}{-\sqrt{1-4y_n}}\, dy_n \qquad \text{(A.10)}$$

$$= 2 \cdot \int_0^{\frac{1}{4}} \delta(x_{n+1} - 4y_n) \cdot \frac{1}{\pi\sqrt{y_n}} \cdot \frac{1}{\sqrt{1-4y_n}}\, dy_n \quad ,$$

where in the last step we exchanged the integration boundaries, eliminating the minus sign in the second integral. Here it becomes clear how important the integration boundaries are in the inversion of the dynamics f in the delta-function. With the next substitution $y_n := 4y_n$ we finally can integrate over the delta-function and obtain

$$p^*(x_{n+1}) = 2 \cdot \int_0^1 \delta(x_{n+1} - z_n) \cdot \frac{1}{\pi\sqrt{\frac{1}{4}z_n}} \cdot \frac{1}{\sqrt{1-z_n}} \cdot \frac{1}{4}\, dz_n$$

$$= \frac{1}{\pi\sqrt{x_{n+1}(1-x_{n+1})}} \quad , \qquad \text{(A.11)}$$

hence the distribution (A.6) is really an invariant distribution of the Ulam map.

A.2 Conjugation of Ulam Map and Tent Map

The analytic form of the stationary distribution was originally obtained via the conjugation $h(x_n) = y_n$ of the Ulam map with the much simpler tractable tent map

$$y_{n+1} := g(y_n) = \begin{cases} 2y_n & \text{for } y_n \in \left[0, \frac{1}{2}\right] \\ 2(1-y_n) & \text{for } y_n \in \left[\frac{1}{2}, 1\right] \end{cases} \quad . \qquad \text{(A.12)}$$

It is easy to show that the invariant density of the tent map is $p_g^*(y_n) = 1$, hence the uniform distribution on the unit interval. Again the solution can be obtained via the solution of the Perron–Frobenius equation for the two branches of the map.

The conjugation is now given by $x_n := \sin^2\left(\frac{\pi}{2}y_n\right)$ [Grossmann, Thomae (1977)] or

$$h(y_n) = \frac{2}{\pi}\arcsin(\sqrt{x_n}) \quad . \tag{A.13}$$

First, one has to show that this really gives a conjugation by fulfiling for both branches each

$$g(y_n) = h \circ f \circ h^{-1}(y_n) \quad . \tag{A.14}$$

For the branch $y_n \in \left[0, \frac{1}{2}\right]$ it is

$$\begin{aligned} g(y_n) &= h \circ f \circ h^{-1}(y_n) \\ &= h \circ f(\sin^2\left(\tfrac{\pi}{2}y_n\right)) \\ &= h(4\sin^2\left(\tfrac{\pi}{2}y_n\right) \cdot (1 - \sin^2\left(\tfrac{\pi}{2}y_n\right))) \\ &= h(4\sin^2\left(\tfrac{\pi}{2}y_n\right) \cdot \cos^2\left(\tfrac{\pi}{2}y_n\right)) \\ &= h(\sin^2(\pi y_n)) \\ &= \tfrac{2}{\pi}\arcsin(\sqrt{\sin^2(\pi y_n)}) \\ &= 2y_n \quad , \end{aligned} \tag{A.15}$$

where for the simplification of the expressions including the angular functions the theorem of Pythagoras (ca. 571–497 b.C.) $1 = \sin^2 x + \cos^2 x$ and the addition theorem $\sin(2x) = 2\sin x \cos x$ have been applied.

For the other branch the calculation can be performed analogously using $\sin\left(\frac{\pi}{2}y\right) = \cos\left(\frac{\pi}{2}(1-y)\right)$ and $\cos\left(\frac{\pi}{2}y\right) = \sin\left(\frac{\pi}{2}(1-y)\right)$ giving

$$\begin{aligned} g(y_n) &= h(4\sin^2\left(\tfrac{\pi}{2}y\right) \cdot \cos^2\left(\tfrac{\pi}{2}y\right)) \\ &= h(4\cos^2\left(\tfrac{\pi}{2}(1-y)\right) \cdot \sin^2\left(\tfrac{\pi}{2}(1-y)\right)) \\ &= 2(1 - y_n) \quad . \end{aligned} \tag{A.16}$$

Hence the map h is really a conjugation between the Ulam map and the tent map.

The invariant density is transformed via the conjugation to

$$p_f^*(x_n)\;dx_n = p_g^*(y_n)\;dy_n \tag{A.17}$$

and with the derivative $\arcsin'(x) = \frac{1}{\sqrt{(1-x)}}$ it is

$$\frac{dy_n}{dx_n} = \frac{d}{dx_n}\left(\frac{2}{\pi}\arcsin(\sqrt{x_n})\right) = \frac{1}{\pi\sqrt{x_n(1-x_n)}} \quad ,$$

thus

$$p_f^*(x_n) = \underbrace{p_g^*(y_n)}_{=1} \cdot \frac{dy_n}{dx_n} = \frac{1}{\pi\sqrt{x_n(1-x_n)}} \quad . \tag{A.18}$$

A.3 Exponential Divergence in the Ulam Map

For the one-dimensional dynamics the definition of Lyapunov exponents is given in time average starting at x_0

$$\lambda(x_0) = \lim_{n\to\infty}\left(\frac{1}{N}\sum_{n=1}^{N}\ln\left|\frac{df}{dx}(x_n)\right|\right) \quad , \tag{A.19}$$

and for ergodic systems, i.e. where a natural measure is present and not partitioned into pieces, it is equal to the ensemble mean

$$\lambda = \int_0^1 \ln\left|\frac{df}{dx}(x)\right| \cdot p^*(x)\;dx \quad . \tag{A.20}$$

Hence for the tent map it is

$$\begin{aligned}
\lambda &= \int_0^1 \ln\left|\tfrac{dg}{dy}(y)\right| \cdot p_g^*(y)\;dy \\
&= \int_0^{\frac{1}{2}} \ln|2| \cdot 1\;dy + \int_{\frac{1}{2}}^1 \ln|-2| \cdot 1\;dy \\
&= \ln|2|\;(\tfrac{1}{2}+\tfrac{1}{2}) \\
&= \ln(2) \quad .
\end{aligned} \tag{A.21}$$

With the positive Lyapunov exponent $\lambda = \ln(2) > 0$, the tent map is prooven to be chaotic, hence sensitive to small perturbations. With the conjugation h between the tent map and the Ulam map, the Lyapunov exponent of the Ulam map is also given by $\ln(2) > 0$, hence the invariant density is also in this case chaotic.

Of course the Lyapunov exponent can be directly calculated for the Ulam map to see if the Lyapunov exponent is really equal to the one of the tent map. Again the conjugation h helps, this time to tackle the integration boundaries

$$\begin{aligned}\lambda &= \int_0^1 \ln\left|\tfrac{df}{dx}\right| \cdot p_f^*(x)\, dx \\ &= \int_0^1 \ln\left|\tfrac{d}{dy}(h^{-1}(g(y))) \cdot \tfrac{dy}{dx}\right| \cdot \underbrace{p_g^*(y)}_{=1}\, dy\end{aligned} \tag{A.22}$$

with the substitution $y = h(x) = \frac{2}{\pi}\text{arcsin}(\sqrt{x})$ and $f(x) = h^{-1} \circ g \circ h(x) = h^{-1} \circ g(y)$. Furthermore, the Eq. (A.17) holds for invariant density, here preserving the integration boundaries 0 and 1.

Again we divide the integral into the two branches. For $y \in \left[0, \frac{1}{2}\right]$ it is now

$$\begin{aligned}\ln\left|\tfrac{d}{dy}(h^{-1}(g(y))) \cdot \tfrac{dy}{dx}\right| &= \ln\left|\tfrac{d}{dy}\left(\sin^2\left(\tfrac{\pi}{2}(2y)\right)\right) \cdot \tfrac{1}{\left(\frac{dx}{dy}\right)}\right| \\ &= \ln\left|\pi \sin(\pi y) \cdot \tfrac{1}{\left(\frac{\pi}{2}\right)\sin(\pi y)}\right| \\ &= \ln(2)\end{aligned}$$

and for the other branch divides similarly. Since the both integrals contribute each with $\frac{1}{2}$, it really is $\lambda = \ln(2)$, also for the Ulam map. The integration boundary $\frac{1}{2}$ is preserved under variable transformation again via the conjugation.

Appendix B

Parameter Estimation for the Autoregressive AR(1)-Process Gives Least Squares Estimators

For the linear autoregressive process AR(1), as it was described in detail in Chapter 1, we derive the likelihood function, and show that the its maximization gives the well-known least squares estimators. In addition we will briefly describe the Bayesian parameter estimation procedure which also follows from the likelihood function. The results shown here hold for more general linear autoregressive models with Gaussian noise. The techniques used for the AR(1)-process can still be applied using a few more intergals for the extra variables.

For epidemic processes which are described by master equations we will show the application of the principles, which we demonstrated here, in Appendix C in more detail.

B.1 Likelihood and its Maximization

From the defining equation of the AR(1)-process $x_{n+1} = ax_n + b + \sigma\varepsilon_n$ we obtain for $N+1$ measured points $\{x_i\}_{i=0}^N$

$$\underbrace{\begin{pmatrix} x_1 \\ x_2 \\ \\ \vdots \\ \\ x_N \end{pmatrix}}_{=:\underline{y}} = \underbrace{\begin{pmatrix} x_0 & 1 \\ x_1 & 1 \\ \\ \vdots & \vdots \\ \\ x_{N-1} & 1 \end{pmatrix}}_{=:A} \cdot \underbrace{\begin{pmatrix} a \\ b \end{pmatrix}}_{=:\underline{c}} + \sigma \underbrace{\begin{pmatrix} \varepsilon_0 \\ \varepsilon_1 \\ \\ \vdots \\ \\ \varepsilon_{N-1} \end{pmatrix}}_{=:\underline{r}} \tag{B.1}$$

and hence

$$\underline{y} = A \cdot \underline{c} + \underline{r} \quad . \tag{B.2}$$

The measured or observed points $\{x_i\}_{i=0}^{N}$ are known, but not the values of the noise ε_n. The least squares approach to parameter estimation minimizes the quadratic distance $||\underline{r}||^2$, resulting in linear equations containing parameters a and b.

Here we start with Eq. (1.21), and use the fact that the transition probability $p(x_{n+1}|x_n)$ is given as a function of the model parameters a, b and σ. The likelihood L is defined by the joint probability of finding the observation points $x_0, x_1, \ldots, x_N$

$$p(x_0, x_1, \ldots, x_N) = \left(\prod_{i=0}^{N-1} p_{a,b,\sigma}(x_{n+1}|x_n) \right) \cdot p_0(x_0) =: L(a, b, \sigma)$$

as a function of the parameters of the underlying model and is for the AR(1) example

$$L(a, b, \sigma) = \left(\frac{1}{\sigma\sqrt{2\pi}} \right)^N \cdot e^{-\frac{1}{2\sigma^2} \sum_{i=0}^{N-1} (x_{i+1} - (ax_i + b))^2} p_0(x_0) \quad . \tag{B.3}$$

Maximization of $\ell := ln(L)$, hence $\partial\ell/\partial a = 0$ gives for values $\hat{a}$ and $\hat{b}$ maximizing L

$$\hat{a} \cdot \sum_{i=0}^{N-1} x_i^2 + \hat{b} \cdot \sum_{i=0}^{N-1} x_i = \sum_{i=0}^{N-1} x_i x_{i+1} \tag{B.4}$$

respectively, $\partial\ell/\partial b = 0$ gives

$$\hat{a} \cdot \sum_{i=0}^{N-1} x_i + \hat{b} \cdot N = \sum_{i=0}^{N-1} x_{i+1} \quad . \tag{B.5}$$

Thus the maximizing value for the parameter σ, i.e. $\hat{\sigma}$ does not enter into these two equations, and so in this model we do not need to calculate $\partial\ell/\partial\sigma = 0$ when we are just interested in estimates of the parameters of the deterministic model a and b. In more general models, as many equations as parameters are needed to obtain the maximum likelihood estimates.

The Eqs. (B.4) and (B.5) are simply the equations for a least squares problem of interpolation. Hence, the least squares procedure implicitly assumes Gaussian noise. They also can be nicely put into matrix form in the following way, so that

$$H\underline{c} = G$$

with $\underline{c} := (\hat{a}, \hat{b})^{tr}$ or as it is often written

$$A^{tr} A\underline{c} = A^{tr}\underline{y} \quad .$$

Here, A is the design matrix given by

$$A := \begin{pmatrix} x_0 & 1 \\ x_1 & 1 \\ \vdots & \vdots \\ x_{N-1} & 1 \end{pmatrix} \tag{B.6}$$

and the covariance matrix $H := A^{tr} A$ is

$$H = \begin{pmatrix} \sum_{i=0}^{N-1} x_i^2 & \sum_{i=0}^{N-1} x_i \\ \sum_{i=0}^{N-1} x_i & \sum_{i=0}^{N-1} 1 \end{pmatrix} \quad . \tag{B.7}$$

In detail

$$A^{tr} A = \begin{pmatrix} x_0 & x_1 & \dots & x_{N-1} \\ 1 & 1 & \dots & 1 \end{pmatrix} \cdot \begin{pmatrix} x_0 & 1 \\ x_1 & 1 \\ \vdots & \vdots \\ x_{N-1} & 1 \end{pmatrix} \tag{B.8}$$

$$= \begin{pmatrix} \sum_{i=0}^{N-1} x_i^2 & \sum_{i=0}^{N-1} x_i \\ \sum_{i=0}^{N-1} x_i & \sum_{i=0}^{N-1} 1 \end{pmatrix} \quad .$$

Furthermore, the target vector $\underline{y}$ is given by

$$\underline{y} := \begin{pmatrix} x_1 \\ x_2 \\ \vdots \\ x_N \end{pmatrix} \tag{B.9}$$

and $G := A^{tr}\underline{y}$, hence

$$G = \begin{pmatrix} \sum_{i=0}^{N-1} x_i x_{i+1} \\ \sum_{i=0}^{N-1} x_{i+1} \end{pmatrix} \quad . \tag{B.10}$$

Finally, with

$$\omega = \sum_{i=0}^{N-1} x_{i+1}^2 \tag{B.11}$$

we obtain for the exponent in the likelihood function

$$\sum_{i=0}^{N-1} (x_{i+1} - (ax_i + b))^2 = \underline{c}^{tr} H \underline{c} - 2\underline{c}^{tr} G + \omega \quad . \tag{B.12}$$

From this process we simulate one realization of a transient time series $\underline{x} := (x_0, x_1, \ldots, x_N)^{tr}$, $\underline{x} \in \mathbb{R}^{N+1}$, with initial condition $x_0 := 0$. Since we choose the parameters to be $a = 0.9$, $b = 0.1$ and $\sigma = 0.2$, the deterministic map (hence $\sigma = 0$) has the fixed point $x_F = 1$ (see Fig. 1.3). From the frequentists' approach of maximum likelihood we obtain the estimates: $\hat{a} = 0.9017$, $\hat{b} = 0.1037$, $\hat{\sigma} = 0.02028$ for this one realization.

From the likelihood function we have calculated the estimates as the maximum of that likelihood, but have not yet quantified how accurate these estimates are, i.e. whether similar parameter values are much less likely to occur or have similar likelihood. In the frequentists' approach to parameter estimation one could look at the curvature of the likelihood function around the maximum to obtain some information of the insecurity about the parameter estimation.

More desirable, however, would be to calculate a properly normalizable distribution of the parameter values after taking the data points into account. The likelihood function as given above does not give any well-behaved, i.e. normalizable, distribution for the parameter values. We will only be able to obtain this in the Bayesian framework of parameter estimation as a posterior distribution of the parameters conditioned on the data, at the cost of having to introduce a well-behaved prior distribution of parameters.

In this Bayesian framework the problem of non-normalizable likelihood functions can be understood as having implicitly used a uniform prior, which on an infinite interval is also infinite.

In the next section, we will introduce the Bayesian approach of parameter estimation in application to our example of the AR(1)-process, where again most calculations can be done explicitly. Only the last step of the Gibbs sampler will be done with computer assistance to obtain a joint posterior for all parameters of the AR(1)-model.

B.2 Bayesian Parameter Estimation

With data $\underline{x}$, and parameters $\underline{\vartheta} := (a, b, \sigma)^{tr}$ respectively $\underline{c} := (a, b)^{tr}$ the Bayesian ansatz reads

$$p(\underline{\vartheta}|\underline{x}) = p(\underline{x}|\underline{\vartheta}) \cdot \frac{p(\underline{\vartheta})}{p(\underline{x})} \tag{B.13}$$

with prior $p(\underline{\vartheta})$, normalization constant

$$p(\underline{x}) = \int p(\underline{x}|\underline{\vartheta})p(\underline{\vartheta})\, d\vartheta \quad =: k$$

and likelihood function

$$p(\underline{x}|\underline{\vartheta}) = \left(\frac{1}{\sigma\sqrt{2\pi}}\right)^N e^{-\frac{1}{2\sigma^2}R(\underline{c})} p_0(x_0)$$

with A^+ the Penrose pseudo inverse, giving finally the joint posterior $p(\underline{\vartheta}|\underline{x})$. It is

$$R = \sum_{i=0}^{N-1} (x_{i+1} - (ax_i + b))^2 = (A\underline{c} - \underline{y})^{tr} =: R(\underline{c}) \tag{B.14}$$

or

$$R(\underset{\sim}{c}) = (\underset{\sim}{c} - A^+\underset{\sim}{y})^{tr} A^{tr} A(\underset{\sim}{c} - A^+\underset{\sim}{y}) \quad . \tag{B.15}$$

For the given likelihood function the single parameter priors $p(\sigma)$ and $p(\underset{\sim}{c})$ can be given in "conjugate priors" form, giving single parameter posteriors $p(\sigma|\underset{\sim}{c})$ and $p(\underset{\sim}{c}|\sigma)$, depending on the other priors. A possible conjugate prior for the joint posterior $p(\sigma, \underset{\sim}{c}|\underset{\sim}{x})$ is often not possible, but has to be numerically calculated using a Monte Carlo Markov Chain, the Gibbs sampler, as demonstrated below.

B.2.1 *Conditional posteriors*

The inverse Γ-distribution is the conjugate prior for σ of the above likelihood function, hence $p(1/\sigma^2)$ is Γ-distributed, and from that

$$p(\sigma) = p_\Gamma\left(\frac{1}{\sigma^2}\right) \cdot \left|\frac{d\sigma^{-2}}{d\sigma}\right| = \frac{2\beta^\alpha}{\Gamma(\alpha)} \sigma^{-2\alpha-1} e^{-\beta/\sigma^2}$$

for prior parameters α and β to be fixed beforehand. Here the gamma function $\Gamma(a) = \int_0^\infty e^{-x} x^{a-1}\, dt$ appears as normalization constant.

The conditional posterior

$$p(\sigma|\underset{\sim}{c}, \underset{\sim}{x}) = \frac{1}{k_\sigma} p(\underset{\sim}{x}|\underset{\sim}{\vartheta}) p(\sigma)$$

with normalization

$$k_\sigma := \int_0^\infty p(\underset{\sim}{x}|\underset{\sim}{c}, \sigma) p(\sigma)\, d\sigma$$

gives

$$beginequation p(\sigma|\underset{\sim}{c}, \underset{\sim}{x}) = \frac{2(\beta + R/2)^{\alpha+N/2}}{\Gamma(\alpha + N/2)} \sigma^{-2(\alpha+N/2)-1} e^{-(\beta+R/2)/\sigma^2}$$

See prior and conditional posterior for the parameter σ in Fig. B.1.

Similarly, the conjugate prior for the parameter set $\underset{\sim}{c}$ is Gaussian, as is the conditional posterior $p(\underset{\sim}{c}|\sigma, \underset{\sim}{x})$. With a prior

$$p(\underset{\sim}{c}) = \frac{1}{2\pi\sqrt{\det V}} e^{-\frac{1}{2}(\underset{\sim}{c}-\underset{\sim}{\mu})^{tr} V^{-1} (\underset{\sim}{c}-\underset{\sim}{\mu})}$$

the posterior is given by

$$p(\underset{\sim}{c}|\sigma, \underset{\sim}{x}) = \frac{\sqrt{\det U}}{2\pi} e^{-\frac{1}{2}(\underset{\sim}{c}-\underset{\sim}{w})^{tr} U (\underset{\sim}{c}-\underset{\sim}{w})}$$

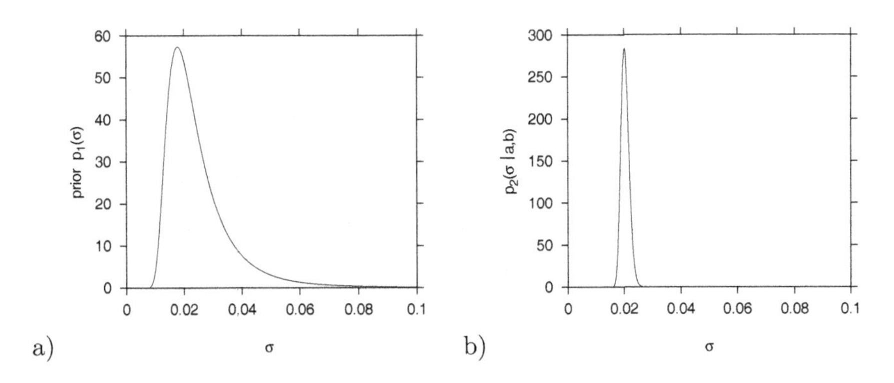

Fig. B.1 AR model, a) Prior and b) posterior, using the previously generated data set from the AR(1)-process, for the parameter σ. As prior values in the inverse gamma-distribution we use $\alpha = 2.0$ and $\beta = 0.0008$. In the posterior these values become $\alpha + N/2 = 52.0$ and $\beta + R/2 = 0.0216$, hence the prior information α and β is overwritten by the information obtained from the data points, i.e. N and R.

with

$$U = \frac{A^{tr} A}{\sigma^2} + V^{-1}$$

and

$$\underline{w} = \left(\frac{A^{tr} A}{\sigma^2} + V^{-1} \right)^{-1} \left(\frac{1}{\sigma^2} A^{tr} \underline{y} - V^{-1} \underline{\mu} \right) \quad .$$

B.2.2 *Gibbs sampler*

In order to obtain the joint posterior from the conditional posteriors as calculated above we construct a Markov chain à la Metropolis [Green (2001)]

$$\frac{p(\sigma, \underline{c}, t + \Delta t)}{\Delta t} = \sum_{\tilde{\sigma}} w_{\sigma, c|\tilde{\sigma}, c} \; p(\tilde{\sigma}, c, t) + \sum_{\tilde{\underline{c}}} w_{\sigma, c|\sigma, \tilde{c}} \; p(\sigma, \tilde{c}, t)$$

with transition rate

$$w_{\sigma, c|\tilde{\sigma}, c} \propto p(\sigma|\tilde{\sigma}, c) \cdot p(c|\tilde{\sigma}, c)$$

with $p(c|\tilde{\sigma}, c) = 1$ and the other terms alike. Detailed balance guarantees the convergence to the joint posterior $p(\sigma, \underline{c}, t) \to p(\underline{\vartheta}|\underline{x})$ for $t \to \infty$.

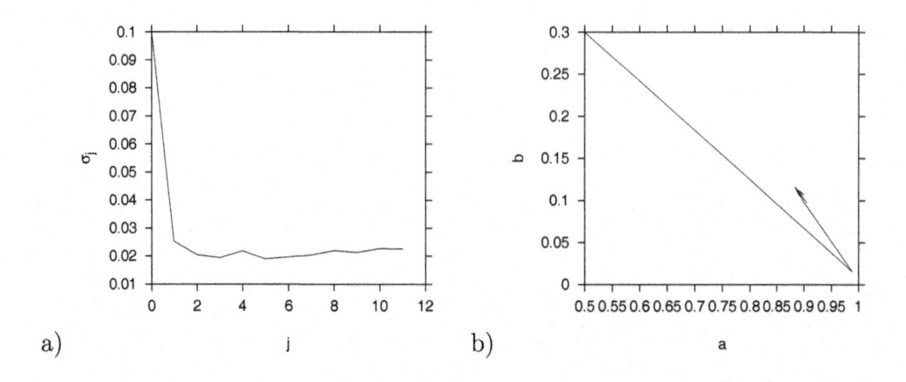

Fig. B.2 AR model, Gibbs sampler. In a) the first iterations are shown for the parameter σ versus iteration steps, in b) the iteration for the two parameters a and b is shown.

The construction of the Markov chain à la Metropolis becomes clearer after looking at magnetic spin systems in Chapter 3, using a dynamic model similar to the Metropolis algorithm. The detailed balance condition is also explained there.

Appendix C

From Stochastic Epidemics to Parameter Estimation

C.1 Likelihood for Simple Stochastic Epidemic Model

We derive the likelihood function for the linear infection model given in Eq. (1.31) hence the master equation becomes

$$\frac{dp(I,t)}{dt} = \sum_{\tilde{I} \neq I} w_{I,\tilde{I}} \; p(\tilde{I}) - \sum_{\tilde{I} \neq I} w_{\tilde{I},I} \; p(I) \tag{C.1}$$

with transition rates for $\underline{n} = (S, I)$

$$\begin{aligned} w_{(S-1,I+1),(S,I)} &= \beta \cdot S \\ w_{(S,I,),(S+1,I-1)} &= \beta \cdot (S+1) \end{aligned}$$

or just for the variable I

$$\frac{dp(I,t)}{dt} = \beta(N-(I-1)) \cdot p(I-1,t) - \beta(N-I) \cdot p(I,t) \quad . \tag{C.2}$$

As boundary equations we have for $I = 0$

$$\frac{dp(I=0,t)}{dt} = -\beta N \cdot p(I=0,t) \tag{C.3}$$

and for $I = N$

$$\frac{dp(I=N,t)}{dt} = \beta \cdot p(I=N-1,t) \quad . \tag{C.4}$$

With the methods described in Section 1.5.1 for the Poisson process one also can solve the original linear infection model Eq. (1.32) thus obtaining

$$p(I,t) = \binom{N}{I} \left(e^{-\beta t}\right)^{N-I} \left(1 - e^{-\beta t}\right)^{I} \tag{C.5}$$

which is a binomial distribution for N trials with I successes and success probability $\left(1 - e^{-\beta t}\right)$. In technical terms, the process is also known as linear death process, death of the initially N susceptibles [Bharucha-Reid (1960)].

Now we look at the mean value of the linear infection model

$$\langle I \rangle := \sum_{I=0}^{N} I \cdot p(I,t) \quad . \tag{C.6}$$

hence

$$\langle I \rangle = N\left(1 - e^{-\beta t}\right) \quad . \tag{C.7}$$

or the ODE

$$\frac{d}{dt} \langle I \rangle = \beta \cdot (N - \langle I \rangle) \tag{C.8}$$

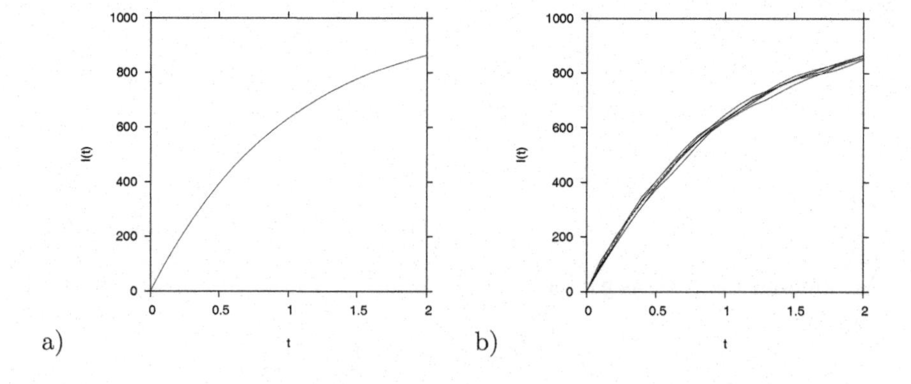

Fig. C.1 a) Mean value dynamics for the simple epidemic process b) Stochastic realizations of the same process compared with the mean values (dashed line).

We show in Fig. C.1a) the solution to Eq. (C.7) e.g. as solution of the ODE, Eq. (C.8), and in C.1b) we show stochastic simulations of the master equation direcly. From these realizations we take one data set of 20 points to estimate the model parameter β when N is already known, hence fixed, and then also estimate simultaneously both free parameters β and N.

C.2 Calculation of the Likelihood Function

To calculate the likelihood function for a stochastic process for $n+1$ data points $I_0(t_0)$, $I_1(t_1)$, $I_2(t_2)$ up to $I_n(t_n)$, we have to calculate the joint probability to finde these data points, given that the model under investigation has parameters, β and N in the present case of the linear infection model. Hence

$$p(I_n, t_n, I_{n-1}, t_{n-1}, ..., I_1, t_1, I_0, t_0) = \prod_{\nu=0}^{n-1} p(I_{\nu+1}, t_{\nu+1} | I_\nu, t_\nu) \cdot p(I_0, t_0) \tag{C.9}$$

with transition probabilities calculated from solving the master equation for the linear infection process which we have given in Eq. (C.2). Hence $p(I_n, t_n, I_{n-1}, t_{n-1}, ..., I_1, t_1, I_0, t_0) =: L(\beta, N)$ will be a function of the model parameters. Using the techniques described in the main text we obtain with initially I_0 already infected at time t_0

$$p(I, t | I_0, t_0) = \binom{N - I_0}{I - I_0} \left(e^{-\beta(t-t_0)}\right)^{N-I} \left(1 - e^{-\beta(t-t_0)}\right)^{I-I_0} \tag{C.10}$$

and obtain for the likelihood function with $\Delta t := (t_{\nu+1} - t_\nu)$

$$L(\beta, N) = \prod_{\nu=0}^{n-1} \binom{N - I_\nu}{I_{\nu+1} - I_\nu} \left(e^{-\beta \Delta t}\right)^{N - I_{\nu+1}} \left(1 - e^{-\beta \Delta t}\right)^{I_{\nu+1} - I_\nu} \tag{C.11}$$

setting $p(I_0, t_0) = 1$ knowing the initial condition from the first observation. We obtain for the logarithm $\ell(\beta, N) := lnL(\beta, N)$ of the likelihood

$$\ell(\beta, N) = \sum_{\nu=0}^{n-1} ln\Gamma(N - I_\nu + 1) - ln\Gamma(N - I_{\nu+1} + 1) - ln\Gamma(I_{\nu+1} - I_\nu + 1)$$
$$-(N - I_{\nu+1})\beta \cdot \Delta t + (I_{\nu+1} - I_\nu) ln\left(1 - e^{-\beta t \Delta}\right)$$

using the Γ-function instead of the factorials in the binomial coefficient $x! = \Gamma(x+1)$ with

$$\Gamma(x) = \int_0^\infty y^{x-1} \, e^{-y} \, dy \tag{C.12}$$

We show the likelihood as function of the two parameters in Fig. C.2, observing a clear maximum for L but with a wide "hill around the top".

From the log-likelihood we calculate the partial derivatives to obtain the maximum and the best estimates as the parameters $\hat{\beta}$ and $\hat{N}$ which

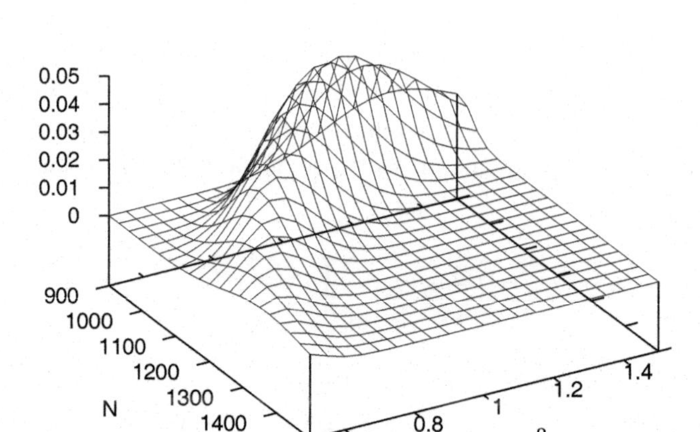

Fig. C.2 Likelihood per data point as function of β and N.

maximise ℓ and so also L. Explicitly, we have with the derivative of the logarithm of the gamma-function given by the digamma-function $\psi(x) := dln(\Gamma(x))/dx$

$$\frac{\partial \ell}{\partial \beta} = \frac{1}{e^{\beta \Delta t} - 1} \left(\sum_{\nu=0}^{n-1} (I_{\nu+1} - I_\nu) \right) \cdot \Delta t - \sum_{\nu=0}^{n-1} (N - I_{\nu+1}) =: f(\beta, N) \quad \text{(C.13)}$$

and

$$\frac{\partial \ell}{\partial N} = \psi(N - I_0 + 1) - \psi(N - I_n + 1) - n\beta\Delta t =: g(\beta, N) \quad . \quad \text{(C.14)}$$

Numerically, the digamma-function is best approximated by

$$\psi(x) = -\gamma - \frac{1}{x} + \sum_{k=1}^{\infty} \frac{x}{k \cdot (x+k)} \quad \text{(C.15)}$$

with Euler's constant $\gamma = \lim_{m \to \infty}((\sum_{\nu=1}^{m} \frac{1}{\nu}) - ln(m))$ which is roughly $\gamma = 0.5772156649$. The maximum of the log-likelihood is given by $f(N, \beta) = 0$ and $g(N, \beta) = 0$ simultaneously. Newton's method applied in two dimensions gives the simultaneous estimates for β and N.

In Fig. C.3 we fix one parameter and show the resulting curve of the first derivative of the maximum likelihood. For instance, fixing $\beta := 1$, we obtain for $g(N)$ the graph in Fig. C.3b) as well for numerically evaluating the derivative of the likelihood as well from the explicit analytical form in Eq. (C.14), giving $g(N)$ since β is fixed. In Fig. C.3a) we fixed $N := 1000$ and show $f(\beta)$ from numeric derivation of the log-likelihood and from the analytic result Eq. (C.13). The zeros in both cases would give estimates for one parameter of the conditioned likelihood, conditioned on the other parameter.

Fixing β gives for the zero of $g(N)$ a value close to the really used value of $N = 1000$, see Fig C.3a). Fixing $N = 1000$ gives for the estimate of β a value of

$$\hat{\beta} = 1.003277 \tag{C.16}$$

for the present realization. Here, the estimate for β ones N is given can be obtained analytically from Eq. (C.13) as

$$\hat{\beta} = \frac{1}{\Delta t} \cdot ln \left(\frac{N - \frac{1}{n} \sum_{\nu=0}^{n-1} I_\nu}{N - \frac{1}{n} \sum_{\nu=0}^{n-1} I_{\nu+1}} \right) \tag{C.17}$$

For the realizations shown in Fig. C.1 the other estimates are $\hat{\beta}_2 = 0.978538$, $\hat{\beta}_3 = 0.959988$, $\hat{\beta}_4 = 0.998381$ and $\hat{\beta}_5 = 0.993164$, just to give a first idea of the variablilty of the estimates depending on the respectively used realization of the stochastic process.

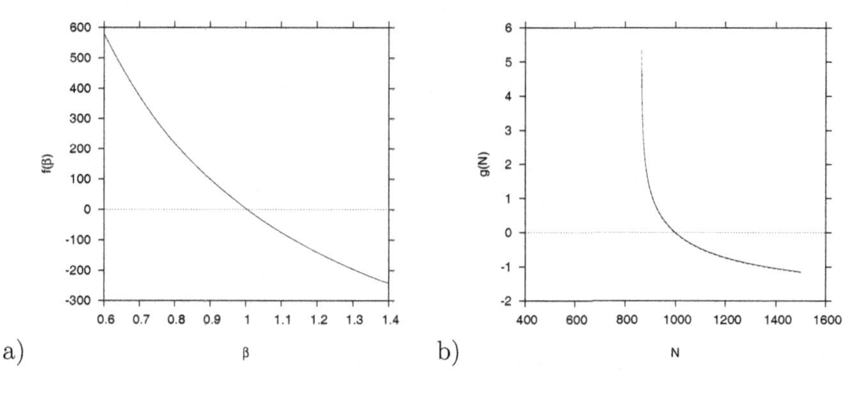

Fig. C.3 a) Plot for $f(\beta)$ from the slope of the likelihood function and from the analytic expression (the curves lie on top of each other). b) Same as in Fig. C.3 a) now for $g(N)$.

We are now interested in the joint estimate of the two only free parameters in our model, β and N. With Newton's method in two dimensions we obtain for our realization of our stochastic process the estimates

$$\hat{\beta} = 1.043 \quad , \quad \hat{N} = 983.7 \quad . \tag{C.18}$$

C.3 Confidence Intervals via Inverse Fisher Matrix

To quantify the insecurity of the estimate, from one realization of the underlying stochastic process we calculate the second derivatives of the log-likelihood function. The negative inverse of this Hessian matrix of the log-likelihood is an estimate of the variance of the likelihood, in case it is approximated by a Gaussian distribution. The negative of the matrix of second derivatives of a log-likelihood is called the Fisher information matrix, in case the expectation value is taken. When only data are available we obtain the observed Fisher matrix to get our estimated confidence intervals.

The second derivatives are given by

$$\begin{aligned}
\frac{\partial^2 \ell}{\partial \beta^2} &= \frac{-(I_n - I_0)(\Delta t)^2 e^{\beta \Delta t}}{\left(e^{\beta \Delta t} - 1\right)^2} \\
\frac{\partial^2 \ell}{\partial N \partial \beta} &= -n \cdot \Delta t \qquad = \frac{\partial^2 \ell}{\partial \beta \partial N} \\
\frac{\partial^2 \ell}{\partial N^2} &= \psi'(N - I_0 + 1) - \psi'(N - I_n + 1)
\end{aligned} \tag{C.19}$$

evaluated at the maximum, hence inserting the estimates $\hat{\beta}$ and $\hat{N}$. Here $\psi'(x)$ is the trigamma-function, the derivative of the digamma-function. It is best approximated for numerical evaluation by

$$\psi'(x) = \sum_{k=0}^{\infty} \frac{1}{(x+k)^2} \tag{C.20}$$

The variance matrix for a two-dimensional Gaussian distribution of the parameters β and N is given by

$$V = -\begin{pmatrix} \frac{\partial^2 \ell}{\partial \beta^2} & \frac{\partial^2 \ell}{\partial N \partial \beta} \\ \frac{\partial^2 \ell}{\partial \beta \partial N} & \frac{\partial^2 \ell}{\partial N^2} \end{pmatrix}^{-1} =: \begin{pmatrix} v_{\beta\beta} & v_{N\beta} \\ v_{\beta N} & v_{NN} \end{pmatrix} \tag{C.21}$$

Hence we assume $L(\beta, N) \approx p(\beta, N)$ with the Gaussian distribution

$$p(\beta, N) = \frac{1}{\sqrt{det(V)} \cdot (\sqrt{2\pi})^2} \; e^{-\frac{1}{2} \begin{pmatrix} \beta - \hat{\beta} \\ N - \hat{N} \end{pmatrix}^{tr} V^{-1} \begin{pmatrix} \beta - \hat{\beta} \\ N - \hat{N} \end{pmatrix}} \tag{C.22}$$

which is a good approximation for a likelihood obtained from many data points.

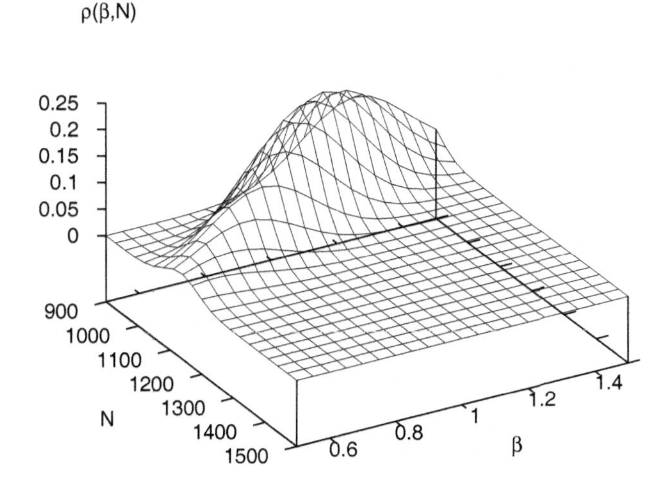

Fig. C.4 The Gaussian distribution per data point $\rho(\beta, N) := p^{1/n}(\beta, N)$ as a function of β and N, as calculated from Eq. (C.22) with the estimated parameter values and confidence intervals.

Finally, the two-sigma confidence interval (which is roughly the 95% confidence interval) is given by twice the square root of the diagonals of the variance matrix, so our estimates finally include the confidence intervals

$$\begin{aligned} \beta &= \hat{\beta} \pm 2 \cdot \sqrt{v_{\beta\beta}} \\ N &= \hat{N} \pm 2 \cdot \sqrt{v_{NN}} \quad . \end{aligned}$$

For the 95% confidence interval the prefactor in front of the square root would be 1.96 instead of the 2.

For our realization of the stochastic process we obtain

$$\beta = 1.043 \pm 2 \cdot 0.064$$
$$N = 983.7 \pm 2 \cdot 21.3$$

as estimates of the two model parameters, or for the two-sigma confidence intervals 0.129 for β and 42.7 for N. The originally used values for generating the data under investigation, $\beta = 1$ and $N = 1000$, lie roughly one standard deviation away from the maximum likelihood estimates.

Fig. C.4 shows the Gaussian approximation to the likelihood function, as obtained from the maximum and the second order approximation to the maximum of the likelihood. It captures well the main shape of the likelihood, but misses the slight curvature of the likelihood "mountain" to a half moon in β-N space.

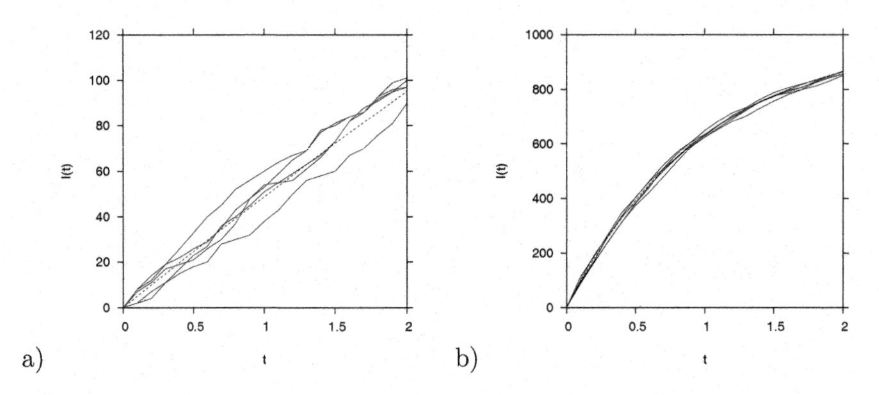

Fig. C.5 a) The relative noise level is larger for smaller β values, here $\beta = 0.05$ as opposed to the original case of b) $\beta = 1$. Also the confidence intervals then become relatively larger to the estimates (for the standard deviation of N we obtain $\sigma_{NN} = 400$, and for β we get $\sigma_{\beta\beta} = 0.06 \approx \hat{\beta}$); the estimated values are less informative.

The likelihood function is a non-normalized function rather than a probability density (and often not even normalizable). To account for this lack of interpretability of the likelihood function as a probability, to obtain the parameters from the data, the Bayesian framework reaches the aim of a posterior distribution of the parameter, given the data in a formal way but on the expense of an ad hoc specified prior distribution. Again, in the case of large data sets, i.e. good information, the prior appears in the posterior only as a slight perturbation, often numerically unimportant.

Analytically, conjugate priors help to obtain closed forms of the posterior distribution of the parameters, but are in many applications difficult to find. For the binomial distribution, the beta-distribution is a conjugate prior, giving a posterior for our parameter β, but still conditioned on N.

C.4 Improving Confidence Intervals

We obtain only symmetric confidence intervals when taking the Fisher matrix approach.

To improve the confidence interval around the estimate for β we investigate further the Bayesian approach, since for the likelihood function as a function of β, a conjugate prior is known, namely the beta distribution.

The likelihood function $L(\beta|N) := L(\beta)$ can be read from Eq. (C.11), namely

$$L(\beta) = \prod_{\nu=0}^{n-1} \binom{N - I_\nu}{I_{\nu+1} - I_\nu} \left(e^{-\beta(\Delta t)}\right)^{N-I_{\nu+1}} \left(1 - e^{-\beta(\Delta t)}\right)^{I_{\nu+1}-I_\nu} \tag{C.23}$$

which is equal to

$$L(p_\beta) = \left(\prod_{\nu=0}^{n-1} \binom{N - I_\nu}{I_{\nu+1} - I_\nu}\right) (1 - p_\beta)^{\sum_{\nu=0}^{n-1}(N-I_{\nu+1})} p_\beta^{\sum_{\nu=0}^{n-1}(I_{\nu+1}-I_\nu)} . \tag{C.24}$$

With the new parameter

$$p_\beta := 1 - e^{-\beta\Delta t} \tag{C.25}$$

and hence with constants

$$k_1 := \left(\prod_{\nu=0}^{n-1} \binom{N - I_\nu}{I_{\nu+1} - I_\nu}\right) \tag{C.26}$$

and

$$k_2 := \sum_{\nu=0}^{n-1} (I_{\nu+1} - I_\nu) \tag{C.27}$$

and

$$k_3 := \sum_{\nu=0}^{n-1} (N - I_{\nu+1}) \tag{C.28}$$

we get

$$L(p_\beta) = k_1 \, p_\beta^{k_2} \, (1 - p_\beta)^{k_3} =: p(\underline{I}|p_\beta) \quad . \tag{C.29}$$

The conjugate prior for this likelihood function is the beta-distribution with parameters a and b

$$p_{a,b}(p_\beta) = \frac{\Gamma(a+b)}{\Gamma(a)\,\Gamma(b)} \, p_\beta^{a-1} \, (1 - p_\beta)^{b-1} =: p(p_\beta) \tag{C.30}$$

with $p(p_\beta) = p(\beta) \frac{d\beta}{dp_\beta}$ from the transformation rules for probability densities.

Now the full Bayesian ansatz is as follows. For the joint probability the order of arguments does not matter, therefore

$$p(\beta, \underline{I}) = p(\underline{I}, \beta) \tag{C.31}$$

and with conditional probabilities

$$p(\beta|\underline{I}) \cdot p(\underline{I}) = p(\underline{I}|\beta) \cdot p(\beta) \tag{C.32}$$

hence for the posterior, the desired probability for the parameter β conditioned on the data observed, it is

$$p(\beta|\underline{I}) = \frac{p(\underline{I}|\beta)}{p(\underline{I})} \, p(\beta) \tag{C.33}$$

with normalization constant

$$p(\underline{I}) = \int p(\underline{I}, \beta) \, d\beta = \int p(\underline{I}|\beta) \cdot p(\beta) \, d\beta \tag{C.34}$$

calculated from the integral over likelihood times prior.

Here we have thus for the likelihood function times the prior

$$p(\underline{I}|p_\beta) \cdot p(p_\beta) = k_1 \, p_\beta^{k_2} \, (1 - p_\beta)^{k_3} \cdot \frac{\Gamma(a+b)}{\Gamma(a)\,\Gamma(b)} \, p_\beta^{a-1} \, (1 - p_\beta)^{b-1} \tag{C.35}$$

which equals with constant

$$k_4 := k_1 \cdot \frac{\Gamma(a+b)}{\Gamma(a)\,\Gamma(b)} \tag{C.36}$$

to

$$p(\underline{I}|p_\beta) \cdot p(p_\beta) = k_4 \, p_\beta^{k_2+a-1} \, (1 - p_\beta)^{k_3+b-1} \tag{C.37}$$

Hence the normalization constant is given by

$$p(\underline{I}) = \int p(\underline{I}, p_\beta)\, dp_\beta = \int_0^1 k_4 p_\beta^{k_2+a-1} (1-p_\beta)^{k_3+b-1}\, dp_\beta \tag{C.38}$$

And with the integral defining the beta-function

$$\int_0^1 x^{a-1}(1-x)^{b-1}\, dx = B(a,b) = \frac{\Gamma(a)\,\Gamma(b)}{\Gamma(a+b)} \tag{C.39}$$

we obtain for the normalization constant

$$p(\underline{I}) = k_4 \frac{\Gamma(a+k_2)\,\Gamma(b+k_3)}{\Gamma(a+k_2+b+k_3)} \tag{C.40}$$

Hence the posterior is given by

$$p(p_\beta|\underline{I}) = \frac{p(\underline{I}|p_\beta)p(p_\beta)}{p(\underline{I})} = \frac{k_4 p_\beta^{k_2+a-1}(1-p_\beta)^{k_3+b-1}}{k_4 \frac{\Gamma(a+k_2)\,\Gamma(b+k_3)}{\Gamma(a+k_2+b+k_3)}} \tag{C.41}$$

which is again a beta distribution, but now for the hyper parameters $(a+k_2)$ and $(b+k_3)$ instead of the prior parameters a and b, explicitly

$$p(p_\beta|\underline{I}) = \frac{\Gamma(a+k_2+b+k_3)}{\Gamma(a+k_2)\,\Gamma(b+k_3)}\, p_\beta^{a+k_2-1}\, (1-p_\beta)^{b+k_3-1} \tag{C.42}$$

In order to give the posterior $p(\beta|\underline{I})$ in original coordinate β instead of p_β we use the definition of $p_\beta := 1 - e^{-\beta\Delta t}$ from Eq. (C.25) and its derivative

$$\frac{dp_\beta}{d\beta} = e^{-\beta\Delta t} \cdot \Delta t \tag{C.43}$$

To transform the probabilities $p(\beta)\, d\beta = p(p_\beta)\, dp_\beta$ hence

$$p(\beta|\underline{I}) = p(p_\beta|\underline{I})\, \frac{dp_\beta}{d\beta} \tag{C.44}$$

and in detail (inserting all previously used abbreviations)

$$\begin{aligned} p(\beta|\underline{I}) = &\frac{\Gamma(a+b+\sum_{\nu=0}^{n-1}(N-I_\nu))}{\Gamma(a+\sum_{\nu=0}^{n-1}(I_{\nu+1}-I_\nu))\,\Gamma(b+\sum_{\nu=0}^{n-1}(N-I_{\nu+1}))} \\ &\cdot \left(1-e^{-\beta\Delta t}\right)^{a+\sum_{\nu=0}^{n-1}(I_{\nu+1}-I_\nu)-1} \left(e^{-\beta\Delta t}\right)^{b+\sum_{\nu=0}^{n-1}(N-I_{\nu+1})-1} \\ &\cdot e^{-\beta\Delta t} \cdot \Delta t \end{aligned} \tag{C.45}$$

Good prior parameters are e.g. $a := 2$ and $b := 2$, whereas the hyper parameters are mainly determinied by the sum over the data points, and so in general have much larger values than the ones for a and b.

For more complicated epidemiological models the calculation effort becomes prohibitive or even impossible. In most cases no analytic results can be given. Hence numerical methods are often required but also are computationally intensive.

Appendix D

Product Space for Spin 1/2 Many-Particle Systems

With single particle creation and annihilation operators

$$c^+ := \begin{pmatrix} 0 & 1 \\ 0 & 0 \end{pmatrix} \quad , \quad c := \begin{pmatrix} 0 & 0 \\ 1 & 0 \end{pmatrix} \quad , \tag{D.1}$$

and single particle state vectors

$$|1\rangle := \begin{pmatrix} 1 \\ 0 \end{pmatrix} \quad , \quad |0\rangle := \begin{pmatrix} 0 \\ 1 \end{pmatrix} \quad , \tag{D.2}$$

the corresponding two-particle system would be constructed as a product space with four-dimensional state vectors and 4×4 matrices via tensor products. Hence the vacuum state is

$$|0\rangle := \begin{pmatrix} 0 \\ 0 \\ 0 \\ 1 \end{pmatrix} \quad , \tag{D.3}$$

and a state containing one particle at site 1 and no particle at site 2, hence the state $|1, 0\rangle$, is

$$|1, 0\rangle = \begin{pmatrix} 0 \\ 1 \\ 0 \\ 0 \end{pmatrix} \quad , \tag{D.4}$$

being created from $c_1^+|0\rangle = c_1^+|0, 0\rangle = |1, 0\rangle$. The creation operators for the two particles are the 4×4 matrices built from 2×2 matrices

$$c_1^+ = \begin{pmatrix} 0 & \mathbb{1} \\ 0 & 0 \end{pmatrix} \quad , \quad c_2^+ = \begin{pmatrix} c^+ & 0 \\ 0 & c^+ \end{pmatrix} \quad , \tag{D.5}$$

with 2×2-unit matrix $\mathbb{1}$ or, written, out e.g.

$$c_1^+ = \begin{pmatrix} 0 & 0 & 1 & 0 \\ 0 & 0 & 0 & 1 \\ 0 & 0 & 0 & 0 \\ 0 & 0 & 0 & 0 \end{pmatrix} \quad . \tag{D.6}$$

The other operators c_1 and c_2 follow directly from this, and commutation rules e.g. $[c_1, c_2^+]_- = 0$ can be shown easily.

The tensor product for vectors is

$$\begin{pmatrix} a_1 \\ a_2 \end{pmatrix} \otimes \begin{pmatrix} b_1 \\ b_2 \end{pmatrix} = \begin{pmatrix} a_1 \cdot \begin{pmatrix} b_1 \\ b_2 \end{pmatrix} \\ \\ a_2 \cdot \begin{pmatrix} b_1 \\ b_2 \end{pmatrix} \end{pmatrix} = \begin{pmatrix} a_1 \cdot b_1 \\ a_1 \cdot b_2 \\ a_2 \cdot b_1 \\ a_2 \cdot b_2 \end{pmatrix} \tag{D.7}$$

respectively for matrices $A \otimes B$

$$\begin{pmatrix} a_{11} & a_{12} \\ a_{21} & a_{22} \end{pmatrix} \otimes B = \begin{pmatrix} a_{11} \cdot B & a_{12} \cdot B \\ a_{21} \cdot B & a_{22} \cdot B \end{pmatrix} \tag{D.8}$$

giving

$$A \otimes B = \begin{pmatrix} a_{11}b_{11} & a_{11}b_{12} & a_{12}b_{11} & a_{12}b_{12} \\ a_{11}b_{21} & a_{11}b_{22} & a_{12}b_{21} & a_{12}b_{22} \\ a_{21}b_{11} & a_{21}b_{12} & a_{22}b_{11} & a_{22}b_{12} \\ a_{21}b_{21} & a_{21}b_{22} & a_{22}b_{21} & a_{22}b_{22} \end{pmatrix} \tag{D.9}$$

as final result. This is also often referred to as the Kronecker product.

Appendix E

Path Integral Using Coherent States for Hard-Core Bosons

The Schrödinger-like equation

$$\frac{d}{dt}|\Psi(t)\rangle = L|\Psi(t)\rangle \tag{E.1}$$

with the Liouville operator

$$L = \sum_{i=1}^{N}(a_i^+ - 1)\beta\left(\sum_{j=1}^{N} J_{ij}\delta_{\hat{n}_j,1}\right)\delta_{\hat{n}_i,0} + \sum_{i=1}^{N}(a_i - 1)\alpha\,\delta_{\hat{n}_i,1} \quad . \tag{E.2}$$

can be integrated formally using $\Delta t \to \infty$ from the quotient of differences

$$\frac{d}{dt}|\Psi(t)\rangle \approx \frac{1}{\Delta t}\left(\,|\Psi(t)\rangle - |\Psi(t-\Delta t)\rangle\,\right) = L(t-\Delta t)\,|\Psi(t-\Delta t)\rangle \tag{E.3}$$

showing

$$|\Psi(t)\rangle = (1 + \Delta t \cdot L(t-\Delta t))\,|\Psi(t-\Delta t)\rangle$$

and for several subsequent time steps

$$|\Psi(t)\rangle = \prod_{\mu=1}^{M}(1 + \Delta t \cdot L(t-\mu\cdot\Delta t))\,|\Psi(t-M\cdot\Delta t)\rangle$$

where $t - M\cdot\Delta t =: t_s$ is the starting time of the stochastic process.

With the Felderhof projection operator [Felderhof (1971)]

$$\langle P| := \langle 0|e^{\sum_{i=1}^{N} a_i} \tag{E.4}$$

and the definition for the state vector

$$|\Psi(t)\rangle = \sum_{I_1=0}^{1} \cdots \sum_{I_N=0}^{1} p(I_1,\ldots,I_N,t)\left(\prod_{i=1}^{N}\left(a_i^+\right)^{I_i}\right)|0\rangle \tag{E.5}$$

any measurable quantity A as a function of the state variables I_i in the master equation formulation with the number operator $a_i^+ a_i$

$$A := A(I_1, ..., I_N) = A(\{I_i\}) = A(\{a_i^+ a_i\}) \tag{E.6}$$

using the notation $\{I_i\} := \{I_1, ..., I_N\}$ has for its expectation value

$$\langle A \rangle(t) := \sum_{\{I_i\}} A(\{I_i\}) p(\{I_i\}, t) \tag{E.7}$$

the following expressions

$$\langle A \rangle(t) = \langle P | A(\{a_i^+, a_i\}) | \Psi(t) \rangle \quad . \tag{E.8}$$

Again we use

$$\sum_{\{I_i\}} := \sum_{I_1=0}^{1} \dots \sum_{I_N=0}^{1} \quad . \tag{E.9}$$

The path integral for an expectation value is then expressed by

$$\langle P | A | \Psi(t_f) \rangle = \langle P | A \prod_{\nu=1}^{M} (1 + \Delta t \cdot L_\nu) | \Psi(t_s) \rangle \tag{E.10}$$

with final time t_f and starting time t_s and times t_ν such that $t_0 = t_f$ and $t_M = t_s$, $L_\nu := L(t_\nu)$.

With coherent states $|\Phi\rangle := e^{\Phi \cdot a^+} |0\rangle$ and its completeness relation

$$\mathbb{1} = \int_{-\infty}^{\infty} \int_{-\infty}^{\infty} \frac{d\Phi^* \, d\Phi}{2\pi i} \; e^{-\Phi^* \Phi} \; |\Phi\rangle \, \langle \Phi| \tag{E.11}$$

and abbreviation $\int d^2\Phi := \int_{-\infty}^{\infty} \int_{-\infty}^{\infty} \frac{d\Phi^* \, d\Phi}{2\pi i}$ we have for N site with operators a_j, a_j^+ the completeness relation

$$\mathbb{1} = \int \left(\prod_{j=1}^{N} d^2\Phi(t_\nu) \right) e^{-\sum_{j=1}^{N} |\Phi_j(t_\nu)|^2} \; |\{\Phi_j(t_\nu)\}_{j=1}^{N}\rangle \, \langle \{\Phi_j(t_\nu)\}_{j=1}^{N}| \tag{E.12}$$

with $|\{\Phi_j(t_\nu)\}_{j=1}^{N}\rangle := |\Phi_1(t_\nu), ..., \Phi_N(t_\nu)\rangle$.

We now can introduce unit operators $\mathbb{1}$ between every time slice of the path integral, and then insert the completeness relations for the coherent states

$$\begin{aligned}
\langle P|A|\Psi(t_f)\rangle &= \langle P|A\, \mathbb{1} \left(\prod_{\nu=1}^{M} (1+\Delta t \cdot L_\nu)\ \mathbb{1} \right) |\Psi(t_s)\rangle \\
&= \int \left(\prod_{j=1}^{N} d^2\Phi(t_f) \right) \langle P|A|\{\Phi_j(t_f)\}_{j=1}^{N}\rangle \\
&\quad \cdot \int \left(\prod_{j=1}^{N} \prod_{\nu=1}^{M} d^2\Phi(t_\nu) \right) \left(\prod_{\nu=1}^{M} e^{-\sum_{j=1}^{N} |\Phi_j(t_{\nu-1})|^2} \right. \\
&\quad \left. \langle \{\Phi_j(t_{\nu-1})\}_{j=1}^{N}|\ (1+\Delta t \cdot L_\nu)|\{\Phi_j(t_\nu)\}_{j=1}^{N}\rangle \right) \\
&\quad \cdot e^{-\sum_{j=1}^{N} |\Phi_j(t_s)|^2}\ \ \langle \{\Phi_j(t_f)\}_{j=1}^{N}|\Psi(t_s)\rangle
\end{aligned} \tag{E.13}$$

considering the non-boundary terms

$$\begin{aligned}
(*) := \int \left(\prod_{j=1}^{N} \prod_{\nu=1}^{M} d^2\Phi(t_\nu) \right) \left(\prod_{\nu=1}^{M} e^{-\sum_{j=1}^{N} |\Phi_j(t_{\nu-1})|^2} \right. \\
\left. \langle \{\Phi_j(t_{\nu-1})\}_{j=1}^{N}|\ (1+\Delta t \cdot L_\nu)|\{\Phi_j(t_\nu)\}_{j=1}^{N}\rangle \right)
\end{aligned} \tag{E.14}$$

further in the following: It is

$$\begin{aligned}
&\langle \{\Phi_j(t_{\nu-1})\}_{j=1}^{N}|\ (1+\Delta t \cdot L_\nu)|\{\Phi_j(t_\nu)\}_{j=1}^{N}\rangle \\
&\qquad = e^{-\sum_{j=1}^{N} \Phi_j^*(t_{\nu-1})\cdot\Phi_j(t_\nu)}\ \ + \Delta t \cdot \tilde{L}_\nu
\end{aligned} \tag{E.15}$$

with

$$\langle \{\Phi_j(t_{\nu-1})\}_{j=1}^{N}|\{\Phi_j(t_\nu)\}_{j=1}^{N}\rangle = e^{-\sum_{j=1}^{N} \Phi_j^*(t_{\nu-1})\cdot\Phi_j(t_\nu)} \tag{E.16}$$

and

$$\tilde{L}_\nu := \langle \{\Phi_j(t_{\nu-1})\}_{j=1}^N | \, L_\nu \, | \{\Phi_j(t_\nu)\}_{j=1}^N \rangle$$

$$= \sum_{k=1}^{N} \alpha \cdot \langle \{\Phi_j(t_{\nu-1})\}_{j=1}^N | \, a_k \, \delta_{\hat{n}_k,1} \, | \{\Phi_j(t_\nu)\}_{j=1}^N \rangle - \alpha \cdot \Phi_k(t_{\nu-1}) \, \Phi_k(t_\nu)$$

$$+ \sum_{k=1}^{N} \beta \cdot \left(\sum_{\ell=1}^{N} J_{k,\ell} \Phi_\ell(t_{\nu-1}) \, \Phi_\ell(t_\nu) e^{\sum_{j=1, j\neq k,\ell}^{N} \Phi_j(t_{\nu-1}) \, \Phi_j(t_\nu)} \right) \tag{E.17}$$

$$\cdot \Big(\Phi_k(t_{\nu-1}) - 1 \Big) \, e^{\sum_{j=1, j\neq k}^{N} \Phi_j(t_{\nu-1}) \, \Phi_j(t_\nu)}$$

with

$$\langle \{\Phi_j(t_{\nu-1})\}_{j=1}^N | \, a_k \, \delta_{\hat{n}_k,1} \, | \{\Phi_j(t_\nu)\}_{j=1}^N \rangle = \Phi_k(t_\nu) \cdot e^{\sum_{j=1, j\neq k}^{N} \Phi_j(t_{\nu-1}) \, \Phi_j(t_\nu)} \tag{E.18}$$

etc. using the coherent state definition

$$| \{\Phi_j(t_\nu)\}_{j=1}^N \rangle := e^{\sum_{j=1}^{N} \Phi_j(t_\nu) \, a_j^+} \, |0\rangle \quad . \tag{E.19}$$

In this way we obtain completely the path integral as given in Eqs. (5.58) to (5.60).

Appendix F

Analytical Power Laws in the Meningitis Model

F.1 Distribution of Total Number of Cases

To find the probability distribution of the number of meningitis cases in the $SIRYX$ model following the introduction of a single mutant-carrying host Y, we consider the master equation for the variables Y and X, still assuming stationarity of the underlying SIR system, i.e. the distribution $p(Y, X, t)$ for $Y = 0$, starting with one mutant Y. $p(Y = 0, X, t)$ gives the distribution of meningitis cases X when the epidemic has died out, meaning $Y = 0$.

To obtain $p(Y = 0, X, t)$ we do not necessarily have to consider the exponential waiting times between events, but only the number of events until the mutants vanish. Hence we consider the following time evolution equation (time-discrete Markov process) for events like the creation of new mutants from those already existing, recovery from mutants and creation of actual meningitis cases from mutant infected:

$$\begin{aligned} p(Y, X, \tau + 1) = \tilde{b} \cdot p(Y - 1, X, \tau) + \tilde{a} \cdot p(Y + 1, X, \tau) \\ + \tilde{g} \cdot p(Y, X - 1, \tau) \end{aligned} \tag{F.1}$$

for discrete times steps τ at which events happen and the parameters

$$\tilde{a} = \frac{a}{a + b + g} = \frac{1}{2} \quad , \quad \tilde{b} = \frac{b}{a + b + g} = \frac{1}{2} - \tilde{g} \quad , \quad \tilde{g} = \frac{g}{a + b + g} = \frac{\varepsilon}{2\beta} \quad . \tag{F.2}$$

The final values, e.g. $\tilde{a} = 1/2$, are obtained by using $g = a - b$ from its definition, see Eq. (8.11), and with $\tilde{g}$ being small and proportional to ε and $\tilde{b}$ only slightly smaller than $\tilde{a}$.

With the boundary equation for the absorbing state $Y = 0$

$$p(Y = 0, X, \tau + 1) = \tilde{a} \cdot p(Y + 1 = 1, X, \tau) + p(Y = 0, X, \tau) \qquad \text{(F.3)}$$

while for $Y = 1$

$$p(Y = 1, X, \tau + 1) = \tilde{a} \cdot p(Y + 1 = 2, X, \tau) + \tilde{g} \cdot p(Y = 1, X - 1, \tau) \quad \text{(F.4)}$$

and for initial condition

$$p(Y, X, \tau = 0) = \delta_{Y,1} \cdot \delta_{X,0} \qquad \text{(F.5)}$$

the dynamic is completely defined. Here we used the Kronecker δ, meaning $\delta_{m,n} = 1$ for $m = n$, else zero.

The solution (see Section F.3) of the distribution of the size of the epidemic after the last host Y carrying the mutant strain has vanished, is given by

$$p(Y = 0, X, \tau) = \tilde{g}^X \cdot \tilde{a} \sum_{\omega=0}^{\omega_{max}} \kappa_{X,\omega} \cdot (\tilde{a} \cdot \tilde{b})^{\omega} \qquad \text{(F.6)}$$

which is essentially a polynomial in the transition probabilities $\tilde{a} \cdot \tilde{b}$, reflecting the random walk in the birth–death process for creating Y cases, and X times the transition $\tilde{g}$ creating disease cases X and one additional transition $\tilde{a}$ to the absorbing state $Y = 0$. The coefficients $\kappa_{\tilde{\gamma},\omega}$ are calculated in Section F.3 as

$$\kappa_{\tilde{\gamma},\omega} = C_{\omega} \cdot \binom{2\mu + X}{X} \qquad \text{(F.7)}$$

with the Catalan numbers $C_{\omega} := \frac{1}{\omega+1} \binom{2\omega}{\omega}$ and ω_{max} given by

$$\omega_{max} = \frac{1}{2}\left(\tau - (X+1) - \begin{Bmatrix} 1 \\ 0 \end{Bmatrix}\right) =: \left\lfloor \frac{1}{2}\left(\tau - (X+1)\right) \right\rfloor \quad , \qquad \text{(F.8)}$$

where $\omega_{max} = \omega_{max}(\tau, X)$ is a function of time τ and size of the epidemics X. The expression $\left\{ \begin{smallmatrix} 1 \\ 0 \end{smallmatrix} \right\}$ means that either 0 or 1 has to be chosen to obtain integer ω, giving the same result as the floor symbol in the rightmost expression. From Section F.3 the more general solution for $p(Y, X, \tau)$ for any Y can also be derived.

F.2 Scaling

To obtain the size of the epidemic in the limit of time τ going to infinity and large sizes of the epidemic we analyze further the size distribution

$$p(X) := \lim_{\tau \to \infty} p(Y=0, X, \tau) = \lim_{\tau \to \infty} \tilde{g}^X \cdot \tilde{a} \sum_{\omega=0}^{\omega_{max}(\tau)} \kappa_{X,\omega} \cdot (\tilde{a}\,\tilde{b})^{\omega} \quad . \tag{F.9}$$

with ω_{max} also being dependent on state X.

For time going to infinity, when the epidemic has almost surely died out, we obtain

$$\lim_{\tau \to \infty} \omega_{max}(\tau) = \infty \tag{F.10}$$

Hence we have

$$p(X) = \tilde{g}^X \cdot \tilde{a} \sum_{\omega=0}^{\infty} \kappa_{X,\mu} \cdot (\tilde{a}\,\tilde{b})^{\omega} \quad . \tag{F.11}$$

It can be shown (see Section F.4) that this is equal to

$$p(X) = \tilde{g}^X \cdot \tilde{a} \cdot \ {}_2F_1\left(\frac{X+1}{2}, \frac{X+2}{2}; 2; 4\tilde{a}\tilde{b}\right) \tag{F.12}$$

where the hypergeometric function is given by

$${}_2F_1(u,v;w;x) = \frac{\Gamma(w)}{\Gamma(u)\Gamma(v)} \sum_{\nu=0}^{\infty} \frac{\Gamma(u+\nu)\Gamma(v+\nu)}{\Gamma(w+\nu)} \cdot \frac{x^{\nu}}{\nu!} \quad . \tag{F.13}$$

Using the definitions for $\tilde{a}$, $\tilde{b}$ and $\tilde{g}$ we obtain for the argument of the hypergeometric function

$$4\tilde{a}\tilde{b} = 1 - 2\tilde{g} = 1 - \frac{\varepsilon}{\beta} = 1 - \eta \tag{F.14}$$

and define $\eta := \frac{\varepsilon}{\beta}$ as small when ε is small. Using known properties of the hypergeometric functions we finally obtain the solution (see Section F.4)

$$p_\eta(X) = \sqrt{\eta} \cdot 2^{-(X+1)} \cdot \ {}_2F_1\left(\frac{3-X}{2}, \frac{2-X}{2}; 2; 1-\eta\right) \tag{F.15}$$

as a function of the parameter η which is proportional to ε, hence becomes small for small pathogenicity rates.

For η to zero, $p_\eta(X)$ vanishes for $X \geq 1$ whereas $p_\eta(X=0)$ goes to 1, hence taking all probability. To better understand the limiting behavior we now consider another quantity, the conditional probability $p_\eta(X|X \geq 1)$ given that there is at least one disease case X.

It turns out (see Section F.4) that

$$p_\eta(X = 0) = \frac{1}{1+\eta} \quad . \tag{F.16}$$

Hence for $X \geq 1$

$$\begin{aligned} p_\eta(X|X \geq 1) &:= \frac{p_\eta(X)}{1 - p_\eta(X = 0)} \\ &= (1 + \sqrt{\eta}) \cdot 2^{-(X+1)} \cdot {}_2F_1\left(\frac{3-X}{2}, \frac{2-X}{2}; 2; 1-\eta\right) \quad . \end{aligned} \tag{F.17}$$

In the limit η to zero we find

$$p_\eta(X|X \geq 1) \to \frac{\Gamma(X - \frac{1}{2})}{2\sqrt{\pi} \cdot \Gamma(1 + X)} \tag{F.18}$$

see Eq. (F.63) in Section F.4 and for large X see Eq. (F.64) in Section F.4

$$p_\eta(X|X \geq 1) \sim \frac{1}{2\sqrt{\pi}} \cdot X^{-\frac{3}{2}} \quad . \tag{F.19}$$

In the same way we find

$$p_\eta(X) \sim \frac{1}{2\sqrt{\pi}} \cdot \eta^{\frac{1}{2}} \cdot X^{-\frac{3}{2}} \tag{F.20}$$

for time $\tau \to \infty$, parameter $\eta \to 0$ and large number of disease cases X.

Hence in total we find the following scaling laws for the distribution of the epidemics

$$p_\eta(X) \sim \eta^{\frac{1}{2}} \tag{F.21}$$

and

$$p_\eta(X) \sim X^{-\frac{3}{2}} \tag{F.22}$$

with critical exponents (of mean field type) $\frac{1}{2}$ and $-\frac{3}{2}$ near the critical value $\varepsilon = 0$, or equivalently $\eta = 0$. For the conditional probability $p_\eta(X|X \geq 1)$ we simply get

$$p_\eta(X|X \geq 1) \sim X^{-\frac{3}{2}} \tag{F.23}$$

independent of any parameter dependence for η. The exponent $-\frac{3}{2}$ is exactly the one used for critical branching processes [Harris (1989); De Los Rios (2001)], which is proven by asymptotics of characteristic functions. In total we have obtained power law behavior for the total size distribution for our simplified YX-model in the limit of vanishing or small pathogenicity.

F.3 Solution of Size Distribution of the Epidemic

In this Appendix we analyze the stochastic dynamic system given by Eqs. (F.1) to (F.5). Since the number of X can only increase, we can easily solve the dynamic for the marginal distribution

$$p(Y,\tau) := \sum_{X=0}^{\infty} p(Y,X,\tau) \tag{F.24}$$

and still extract the number of cases X after the epidemic has finished.

Hence the dynamic is

$$p(Y,\tau+1) = \tilde{b}\cdot p(Y-1,\tau) + \tilde{a}\cdot p(Y+1,\tau) + \tilde{g}\cdot p(Y,\tau) \tag{F.25}$$

with boundary

$$p(Y=0,\tau+1) = \tilde{a}\cdot p(Y+1=1,\tau) + p(Y=0,\tau) \tag{F.26}$$

and

$$p(Y=1,\tau+1) = \tilde{a}\cdot p(Y+1=2,\tau) + \tilde{g}\cdot p(Y=1,\tau) \tag{F.27}$$

and initial distribution

$$p(Y,\tau=0) = \delta_{Y,1} \quad . \tag{F.28}$$

With the initial condition vector $(p(Y=0,\tau=0), p(Y=1,\tau=0), p(Y=2,\tau=0), p(Y=3,\tau=0), ...)^{tr} = (0,1,0,0,...)^{tr}$, for time step τ this is a $(\tau+1)$-dimensional matrix system of equations

$$\begin{pmatrix} p(Y=0,\tau+1) \\ p(Y=1,\tau+1) \\ p(Y=2,\tau+1) \\ \vdots \\ p(Y=\tau+1,\tau+1) \end{pmatrix} = \begin{pmatrix} 1 & \tilde{a} & 0 & 0 & 0 \\ 0 & \tilde{g} & \tilde{a} & 0 & 0 \\ 0 & \tilde{b} & \tilde{g} & \tilde{a} & 0 \\ & & \ddots & \ddots & \ddots \\ 0 & 0 & 0 & \tilde{b} & \tilde{g} \end{pmatrix} \cdot \begin{pmatrix} p(Y=0,\tau) \\ p(Y=1,\tau) \\ p(Y=2,\tau) \\ \vdots \\ p(Y=\tau+1,\tau) \end{pmatrix} \tag{F.29}$$

with tridiagonal structure.

The solution of the matrix system is for times $\tau = 1, 2, ...$ and for states $Y = 1, ..., \tau + 1$ given by

$$p(Y, \tau) = \sum_{\tilde{\alpha}=0}^{\tilde{\alpha}_{max}} k_{\tau,Y,\tilde{\alpha}} \cdot \tilde{a}^{\tilde{\alpha}} \, \tilde{b}^{\tilde{\beta}} \, \tilde{g}^{\tilde{\gamma}} \tag{F.30}$$

with

$$\tilde{\beta} = Y - 1 + \tilde{\alpha} \tag{F.31}$$

$$\tilde{\gamma} = \tau - (Y - 1) - 2\tilde{\alpha} \tag{F.32}$$

since it has to be $\tilde{\alpha}+\tilde{\beta}+\tilde{\gamma} = \tau$ the number of total transitions. Furthermore,

$$\tilde{\alpha}_{max} = \left\lfloor \frac{1}{2} (\tau - (Y - 1)) \right\rfloor \tag{F.33}$$

with 0 or 1 to give an integer $\tilde{\alpha}_{max}$. The coefficients $k_{\tau,Y,\tilde{\alpha}}$ fulfil the recursion

$$k_{\tau+1,Y,\tilde{\alpha}} = k_{\tau,Y-1,\tilde{\alpha}} + k_{\tau,Y,\tilde{\alpha}} + k_{\tau,Y+1,\tilde{\alpha}-1} \tag{F.34}$$

for initially $k_{0,1,0} = 1$ and the other coefficients zero.

The solution of this recursion is given by

$$k_{\tau,Y,\tilde{\alpha}} = \frac{Y \cdot (\tau!)}{(Y + \tilde{\alpha})! \, \tilde{\alpha}! \, (\tau - (Y - 1) - 2\tilde{\alpha})!} \tag{F.35}$$

as can be seen by insertion. In terms of $\tilde{\alpha}$, $\tilde{\beta}$ and $\tilde{\gamma}$ it is

$$k_{\tilde{\alpha},\tilde{\beta},\tilde{\gamma}} = \frac{\binom{\tilde{\beta}}{\tilde{\alpha}}}{\binom{\tilde{\beta}+1}{\tilde{\alpha}}} \quad \cdot \quad \frac{\tau!}{\tilde{\alpha}! \, \tilde{\beta}! \, \tilde{\gamma}!} \quad . \tag{F.36}$$

The general solution $p(Y, X)$ can now be read of from the above by reordering the summations such that we sum up the powers of $\tilde{g}$, hence summing over $\tilde{\gamma}$. We will show this now for the absorbing state $Y = 0$.

The distribution of the absorbing state $Y = 0$ is given from its definition

$$p(Y = 0, \tau) := \sum_{\nu=0}^{\tau-1} \tilde{a} \cdot p(Y + 1 = 1, \nu) \tag{F.37}$$

as

$$p(Y=0,\tau)=\sum_{\nu=0}^{\tau-1}\tilde{a}\cdot\sum_{\tilde{\alpha}=0}^{\tilde{\alpha}_{max}}k_{\nu,1,\tilde{\alpha}}\cdot\tilde{a}^{\tilde{\alpha}}\,\tilde{b}^{\tilde{\alpha}}\,\tilde{g}^{\nu-2\tilde{\alpha}} \tag{F.38}$$

or, labeling the sum in the number of transitions creating X, i.e. in powers of $\tilde{g}$

$$p(Y=0,\tau)=\sum_{\tilde{\gamma}=0}^{\tau-1}\tilde{g}^{\tilde{\gamma}}\cdot\tilde{a}\sum_{\mu=0}^{\mu_{max}}\kappa_{\tilde{\gamma},\mu}\cdot(\tilde{a}\,\tilde{b})^{\mu} \tag{F.39}$$

with

$$\mu_{max}=\left\lfloor\frac{1}{2}\left(\tau-(\tilde{\gamma}+1)-\left\{\begin{matrix}1\\0\end{matrix}\right\}\right)\right\rfloor \tag{F.40}$$

and

$$\kappa_{\tilde{\gamma},\mu}:=k_{\tilde{\gamma}+2\mu,1,\mu}\quad . \tag{F.41}$$

Note that here μ and ν are summation indices and not to be confused with the mutation rates mentioned in the main text. The coefficients $\kappa_{\tilde{\gamma},\mu}$ can be expressed in terms of the Catalan numbers $C_\mu:=\frac{1}{\mu+1}\binom{2\mu}{\mu}$ as

$$\kappa_{\tilde{\gamma},\mu}=C_\mu\cdot\binom{2\mu+\tilde{\gamma}}{\tilde{\gamma}}\quad . \tag{F.42}$$

This completes the calculation of the distribution of the size of the epidemics as

$$p(Y=0,X,\tau)=\tilde{g}^{X}\cdot\tilde{a}\sum_{\mu=0}^{\mu_{max}}\kappa_{X,\mu}\cdot(\tilde{a}\,\tilde{b})^{\mu}\quad . \tag{F.43}$$

F.4 Size Distribution of the Epidemic for ε Zero

For $p(Y=0,X,\tau)$ from section F.3 we consider now the limiting behavior for time τ going to infinity and then large X. As described in the main text we set μ_{max} to infinity for τ going to infinity (also before we used ω_{max} to avoid confusion with other notations).

Using the gamma-function to express the factorials, $\Gamma(x+1) = x!$, the coefficients κ are given by

$$\kappa_{\tilde{\gamma},\mu} = C_\mu \cdot \binom{2\mu + \tilde{\gamma}}{\tilde{\gamma}} = \frac{1}{\Gamma(\tilde{\gamma}+1)} \cdot \frac{\Gamma(\tilde{\gamma}+1+2\mu)}{\Gamma(\mu+2)} \cdot \frac{1}{\mu!} \quad . \tag{F.44}$$

Hence the size distribution of the epidemics is given by

$$p(X) = \frac{\tilde{g}^X \cdot \tilde{a}}{\Gamma(X+1)} \sum_{\mu=0}^{\infty} \frac{\Gamma(X+1+2\mu)}{\Gamma(\mu+2)} \cdot \frac{(\tilde{a}\,\tilde{b})^\mu}{\mu!} \quad . \tag{F.45}$$

Using the duplication formula $\Gamma(2 \cdot x) = \frac{1}{\sqrt{2\pi}} \cdot 2^{2x-\frac{1}{2}} \cdot \Gamma(x)\Gamma(x+\frac{1}{2})$ for the Γ-function we can write

$$p(X) = \frac{\tilde{g}^X \cdot \tilde{a}}{\Gamma(X+1)} \cdot \frac{2^{X+\frac{1}{2}}}{\sqrt{2\pi}} \sum_{\mu=0}^{\infty} \frac{\Gamma(\frac{X+1}{2}+\mu) \cdot \Gamma(\frac{X+2}{2}+\mu)}{\Gamma(\mu+2)} \cdot \frac{(4\,\tilde{a}\,\tilde{b})^\mu}{\mu!} \tag{F.46}$$

in the form of a Gauss hypergeometric function ${}_2F_1$ which is defined as

$${}_2F_1(u,v;w;x) := \sum_{\nu=0}^{\infty} \frac{(u)_\nu \cdot (v)_\nu}{(w)_\nu} \cdot \frac{x^\nu}{\nu!} \tag{F.47}$$

[Abramowitz and Stegun (1972)] and with Pochhammer's symbol defined as $(u)_\nu := \frac{\Gamma(u+\nu)}{\Gamma(u)}$ and $(u)_0 := 1$ resulting in

$${}_2F_1(u,v;w;x) = \frac{\Gamma(w)}{\Gamma(u)\Gamma(v)} \sum_{\nu=0}^{\infty} \frac{\Gamma(u+\nu)\Gamma(v+\nu)}{\Gamma(w+\nu)} \cdot \frac{x^\nu}{\nu!} \quad . \tag{F.48}$$

The hypergeometric function ${}_2F_1(u,v;w;x)$ is the solution of the hypergeometric differential equation

$$x(1-x)\frac{d^2F}{dx^2} = u \cdot v \cdot F - (w - (u+v+1) \cdot x)\frac{dF}{dx} \quad . \tag{F.49}$$

Hence with $u := \frac{X+1}{2}$, $v := \frac{X+2}{2}$ and $w := 2$ we finally get an expression for the distribution of the total size of the epidemics in terms of a hypergeometric function

$$p(X) = \tilde{g}^X \cdot \tilde{a} \cdot \; {}_2F_1\left(\frac{X+1}{2}, \frac{X+2}{2}; 2; 4\tilde{a}\tilde{b}\right) \tag{F.50}$$

again using the duplication formula for the Γ-function.

Using the definitions for $\tilde{a}$, $\tilde{b}$ and $\tilde{g}$ we obtain for the argument of the hypergeometric function

$$4\tilde{a}\tilde{b} = 1 - 2\tilde{g} = 1 - \frac{\varepsilon}{\beta} = 1 - \eta \tag{F.51}$$

and define $\eta := \frac{\varepsilon}{\beta}$ as small when ε is small. For the prefactor in front of the hypergeometric function we then obtain

$$\tilde{g}^X \cdot \tilde{a} = \eta^X \cdot 2^{-(X+1)} \quad . \tag{F.52}$$

Hence

$$p_\eta(X) = \eta^X \cdot 2^{-(X+1)} \cdot \quad {}_2F_1\left(\frac{X+1}{2}, \frac{X+2}{2}; 2; 1-\eta\right) \tag{F.53}$$

which can be simplified further using the formula for the hypergeometric function

$${}_2F_1(u, v; w; x) = (1-x)^{w-u-v} \ {}_2F_1(w-u, w-v; w; x) \quad . \tag{F.54}$$

This results in

$$p_\eta(X) = \sqrt{\eta} \cdot 2^{-(X+1)} \cdot \quad {}_2F_1\left(\frac{3-X}{2}, \frac{2-X}{2}; 2; 1-\eta\right) \quad . \tag{F.55}$$

Now we consider the conditional probability given at least one disease case X, i.e. $p_\eta(X|X \geq 1)$. It is given by Bayes' rule $p(X, X \geq 1) = p(X|X \geq 1) \cdot p(X \geq 1)$ and $p(X \geq 1) = 1 - p(X = 0)$. For all $X \geq 1$ we can use $p(X, X \geq 1) = p(X)$ with $p(X)$ from Eq. (F.15). In total we obtain

$$\begin{aligned} p_\eta(X|X \geq 1) &= \frac{p_\eta(X)}{1 - p_\eta(X=0)} \\ &= \frac{\sqrt{\eta} \cdot 2^{-(X+1)} \cdot \quad {}_2F_1\left(\frac{3-X}{2}, \frac{2-X}{2}; 2; 1-\eta\right)}{1 - p_\eta(X=0)} \quad . \end{aligned} \tag{F.56}$$

It is $p_\eta(X = 0) = \sqrt{\eta} \cdot 2^{-1} \cdot \ {}_2F_1\left(3/2, 1; 2; 1-\eta\right)$. Using the integral representation of the hypergeometric function (see [Abramowitz and Stegun (1972)], p. 558, there Eq. 15.3.1)

$${}_2F_1(u, v; w; x) = \frac{\Gamma(w)}{\Gamma(u-w)\Gamma(v)} \int_0^1 z^{v-1}(1-z)^{w-v-1}(1-z \cdot x)^{-u} \, dz \quad . \tag{F.57}$$

with $u = 3/2$, $v = 1$, $w = 2$ and $x = 1 - \eta$ the integral can be solved analytically with elementary algebra obtaining

$$ {}_2F_1(3/2, 1; 2; 1-\eta) = \int_0^1 (1 - z\cdot(1-\eta))^{-\frac{3}{2}}\, dz = \frac{2}{1-\eta}\cdot\left(\frac{1}{\sqrt{\eta}} - 1\right) \quad \text{(F.58)} $$

So for $p_\eta(X = 0)$ we obtain a very simple expression

$$ p_\eta(X = 0) = \frac{1}{1+\eta} \quad . \quad \text{(F.59)} $$

Inserting this result into Eq. (F.56) gives

$$ p_\eta(X|X \geq 1) = (1 + \sqrt{\eta}) \cdot 2^{-(X+1)} \cdot \ {}_2F_1\left(\frac{3-X}{2}, \frac{2-X}{2}; 2; 1-\eta\right) \quad . \quad \text{(F.60)} $$

As opposed to $p_\eta(X)$ from Eq. (F.15), this expression for $p_\eta(X|X \geq 1)$, Eq. (F.60) gives non-vanishing results in the limit $\eta \to 0$. For $\eta = 0$ the hypergeometric function can be given explicitly by

$$ {}_2F_1(u, v; w; 1) = \frac{\Gamma(w)\Gamma(w-u-v)}{\Gamma(w-u)\Gamma(w-v)} \quad . \quad \text{(F.61)} $$

Hence

$$ \begin{aligned} p_\eta(X|X \geq 1) &= (1 + \sqrt{\eta}) \cdot 2^{-(X+1)} \cdot \ {}_2F_1\left(\frac{3-X}{2}, \frac{2-X}{2}; 2; 1-\eta\right) \quad \text{(F.62)} \\ &\to {}_2F_1\left(\frac{3-X}{2}, \frac{2-X}{2}; 2; 1\right) \\ &= 2^{-(X+1)} \cdot \ \frac{\Gamma(2)\Gamma(x - \frac{1}{2})}{\Gamma(\frac{1+X}{2})\Gamma(\frac{1+X}{2} + \frac{1}{2})} \end{aligned} $$

for $\eta \to 0$. $\Gamma(2) = 1$.

Using the duplication formula for Γ-functions again and Stirling's formula for large X-values in the arguments of the remaining Γ-functions, we obtain

$$
\begin{aligned}
p_\eta(X|X \geq 1) &\to \underbrace{2^{-(X+1)} \cdot \frac{\Gamma(X-\frac{1}{2})}{\Gamma(1+X)\cdot\sqrt{2\pi}\cdot 2^{-(X+1)+\frac{1}{2}}}}_{\text{for } \eta\to 0} \\
&= \frac{\Gamma(X-\frac{1}{2})}{2\sqrt{\pi}\cdot\Gamma(1+X)} \\
&\sim \underbrace{\frac{e^{\frac{3}{2}}}{2\sqrt{\pi}} \cdot \frac{(X-\frac{1}{2})^{(X-\frac{1}{2})}}{(X+1)^{(X+1)}}}_{\text{for } X\to\infty} \quad .
\end{aligned}
\tag{F.63}
$$

Finally, we obtain for the X-dependent part in the limit $X \to \infty$

$$
\frac{(X-\frac{1}{2})^{(X-\frac{1}{2})}}{(X+1)^{(X+1)}} \sim X^{-\frac{3}{2}} \cdot e^{-\frac{3}{2}}
\tag{F.64}
$$

as can be seen by

$$
\begin{aligned}
\ln\left(\frac{(X-\frac{1}{2})^{(X-\frac{1}{2})}}{(X+1)^{(X+1)}}\right) &= \left(X-\frac{1}{2}\right)\cdot\ln\left(X-\frac{1}{2}\right) - (X+1)\cdot\ln\left(X+1\right) \\
&= \left(X-\frac{1}{2}\right)\cdot\ln\left(X\left(1-\frac{1}{2X}\right)\right) \\
&\qquad -(X+1)\cdot\ln\left(X\left(1+\frac{1}{X}\right)\right) \\
&\sim \left(X-\frac{1}{2}\right)\cdot\ln(X) - \frac{1}{2} - \Big((X+1)\cdot\ln(X)+1\Big) \\
&= -\frac{3}{2}\cdot\ln(X) - \frac{3}{2} \quad .
\end{aligned}
\tag{F.65}
$$

We use the Taylor expansion of $\ln(1+y)$ around $y=0$ for $y := (1/x)$ in $(x+a)\ln(x+a) = (x+a)(\ln x + \ln(1+(1/x)))$ which gives $(x+a)\ln x + a + a^2/x$ plus higher order terms.

In mathematical terms we find in summary

$$
\lim_{X\to\infty}\lim_{\eta\to 0}\lim_{\tau\to\infty}\left(\frac{p_\eta(X,\tau)}{\sqrt{\eta}\cdot X^{-\frac{3}{2}}}\right) = \text{const.} = \frac{1}{2\sqrt{\pi}}
\tag{F.66}
$$

respectively for the conditional probability $p_\eta(X|X \geq 1)$

$$\lim_{X\to\infty}\lim_{\eta\to 0}\lim_{\tau\to\infty}\left(\frac{p_\eta(X,\tau|X\geq 1)}{X^{-\frac{3}{2}}}\right) = \text{const.} = \frac{1}{2\sqrt{\pi}} \tag{F.67}$$

where the sequence of the limits taken is important. First we let time τ go to infinity, only leaving considerations for relatively small values of infected X, then look close to the critical value $\eta \geq 0$, taking the divergence of $\sqrt{\eta}$ into account. We finally find scaling with a power law for X only for large values of X, but always much smaller than time τ. The actual value of the constant $\frac{1}{2\sqrt{\pi}}$ is of no further importance, apart from for numerical checks.

Bibliography

Chapter 1

Beck, C. & Schlögl, F. (1993) *Thermodynamics of chaotic systems* (Cambridge University Press, Cambridge).

Bharucha-Reid, A.T. (1960) *Elements of the Theory of Markov Processes and Their Applications* (McGraw-Hill, New York).

Feigenbaum, M.J. (1978) Quantitative universality for a class of nonlinear transformations, *J. Stat. Phys.* **19**, 25–52.

May, R.M. (1974) Biological populations with nonoverlapping generations: stable points, stable cycles, and chaos, *Science* **186**, 645–647.

May, R.M. (1976) Simple mathematical models with very complicated dynamics, *Nature* **261**, 459–467.

Ulam, S.M. & von Neumann, J. (1947) On combinations of stochastic and deterministic processes, *Bull. Am. Math. Soc.* **53**, 1120.

van Kampen, N.G. (1992) *Stochastic Processes in Physics and Chemistry* (North-Holland, Amsterdam).

Chapter 2

Brunel, V., Oerding, K. & Wijland, F. (2000) Fermionic field theory for directed percolation in (1+1)-dimension, *J. Phys.* **A 33**, 1085–1097.

Cardy, J. & Täuber, U.C. (1998) Field theory of branching and annihilating random walks, *J. Stat. Phys.* **90**, 1–56.

Dickman, R. & da Silva, J. K. (1998) Moment ratios for absorbing-state phase transitions, *Phys. Rev.* **E 58**, 4266–4270.

Dieckmann, U., Law, R. & Metz, J.A.J. (2000) *The Geometry of Ecological Interactions* (Cambridge University Press, Cambridge).

Ferguson, N.M., Donnelly, C.A. & Anderson, R.M. (2001) The foot-and-mouth epidemic in Great Britain: Pattern of spread and impact of intervention, *Science* **292**, 1155–1160.

Glauber, R.J. (1963) Time-dependent statistics of the Ising model, *J. Math. Phys.* **4**, 294–307.

Goel, N.S. & Richter-Dyn, N. (1974) *Stochastic Models in Biology* (Academic Press, New York).

Grassberger, P. (1983) On the critical behavior of the general epidemic process and dynamical percolation, *Math. Biosci.* **63**, 157–172.

Grassberger, P. & Scheunert, M. (1980) Fock-space methods for identical classical objects, *Fortschr. Phys.* **28**, 547–578.

Greiner, W., Neise, L. & Stöcker, H. (1987) *Theoretische Physik, Band 9, Thermodynamik und Statistische Mechanik* (Verlag Harri Deutsch, Thun, Frankfurt am Main).

Joo, J. & Lebowitz, J. L. (2004) Pair approximation of the stochastic susceptible-infected-recovered-susceptible epidemic model on the hypercubic lattice, *Phys. Rev.* **E 70**, 036114.

Keeling, M.J., Rand, D.A. & Morris, A.J. (1997) Correlation models for childhood epidemics, *Proc. R. Soc. Lond.* **B 264**, 1149–1156.

Levin, S.A. & Durrett, R. (1996) From individuals to epidemics, *Phil. Trans. R. Soc. Lond.* **B 351**, 1615–1621.

Martins, J., Pinto, A. & Stollenwerk, N. (2009) A scaling analysis in the SIRI epidemiological model, *J. Biol. Dynam.* **3**, 479–496.

Peliti, L. (1985) Path integral approach to birth-death processes on a lattice, *J. Physique* **46**, 1469–1483.

Rand, D.A. (1999) Correlation equations and pair approximations for spatial ecologies. In: *Advanced Ecological Theory*, ed. J. McGlade, (Blackwell Science, Oxford, London, Edinburgh, Paris), 100–142.

Stollenwerk, N., Martins, J. & Pinto, A. (2007) The phase transition lines in pair approximation for the basic reinfection model SIRI, *Phys. Lett. A* **371**, 379–388.

Chapter 3

Binney, J.J., Dowrick, N.J., Fisher, A.J. & Newman, M.E.J. (1992) *The Theory of Critical Phenomena, An Introduction to the Renormalization Group* (Oxford University Press, Oxford).

Ising, E. (1925) Beitrag zur Theorie des Ferromagnetismus. *Z. Phys.* **31**, 253–258.

Landau, D.P. & Binder, K. (2000) *Monte Carlo Simulations in Statistical Physics* (Cambridge University Press, Cambridge).

Le Bellac, M. (1991) *Quantum and Statistical Field Theory* (Oxford University Press, Oxford).

Onsager, L. (1944) A two-dimensional model with order-disorder transition, *Phys. Rev.* **65**, 117–149.

Stanley, H.E. (1971) *An Introduction to Phase Transitions and Critical Phenomena* (Oxford University Press, Oxford).

Stauffer, D. (1997) Relaxation of Ising models near and away from criticality, *Physica A* **244**, 344–357.

Tiggemann, D. (2004) New results for the dynamic critical behaviour of the two-dimensional Ising model, *cond-mat/0404050v1*.

Yeomans, J.M. (1992) *Statistical Mechanics of Phase Transitions* (Oxford University Press, Oxford).

Zinn-Justin, J. (1989) *Quantum Field Theory and critical phenomena* (Oxford University Press, Oxford).

Chapter 4

Breban, R. & Blower, S. (2005) The reinfection threshold does not exist, *J. Theor. Biol.* **235**, 151–152.

Boto, J.P. & Stollenwerk N. (2009) Fractional calculus and Levy flights: modelling spatial epidemic spreading. In *Proceedings of 9th Conference on Computational and Mathematical Methods in Science and Engineering, CMMSE 2009*, ed. J.V. Aguiar *et al.* (Salamanca), 177–188.

Brockmann, D. & Hufnagel, L. (2007) Front propagation in reaction-superdiffusion dynamics: Taming Lévy flights with fluctuations, *Phys. Rev. Lett.* **98**, 178301.

Dammer, S.M. & Hinrichsen, H. (2004) Spreading with immunization in high dimensions, *J. Stat. Mech: Theor Exp.* **P07011** +17.

Dickman, R. & Vidigal, R. (2002) Quasi-stationary distributions for stochastic processes with an absorbing state, *cond-mat/0110557*.

Gillespie, D.T. (1976) A general method for numerically simulating the stochastic time evolution of coupled chemical reactions, *J. Comput. Phys.* **22**, 403–434.

Gillespie, D.T. (1978) Monte Carlo simulation of random walks with residence time dependent transition probability rates, *J. Comput. Phys.* **28**, 395–407.

Gomes, G.M.G., White, L.J. & Medley, G.F. (2004) Infection, reinfection and vaccination under suboptimal protection: epidemiological perspective, *J. Theor. Biol.* **228**, 539–549.

Gomes, G.M.G., White, L.J. & Medley, G.F. (2005) The reinfection threshold, *J. Theor. Biol.* **236**, 111–113.

Grassberger, P. (1983) On the critical behavior of the general epidemic process and dynamical percolation, *Math. Biosci.* **63**, 157–172.

Grassberger, P., Chaté, H. & Rousseau, G. (1997) Spreading in media with long-time memeory, *Phys. Rev.* **E 55**, 2488–2495.

Lübeck, S. & Willmann, R.D. (2002) Universal scaling behaviour of directed percolation and the pair contact process in an external field, *J. Phys. A. Math. Gen.* **35**, 10205.

Chapter 5

Aguado, M., Asorey, M., Ercolessi, E., Ortolani, F. & Pasini, S. (2008) Numerical simulation of the SU(3) AFM Heisenberg model, *arxiv: 0801.3565v1*.

Brunel, V., Oerding, K. & Wijland, F. (2000) Fermionic field theory for directed percolation in (1+1)-dimension, *J. Phys.* **A 33**, 1085–1097.

de Oliveira, M.J. (2006) Perturbation series expansion for the gap of the evolution operator associated with the contact process, *Phys. Rev.* **E 74**, 041121.

Dickman, R. & Jensen, I. (1991) Time-Dependent perturbation theory for nonequilibrium lattice models, *Phys. Rev. Lett.* **67**, 2391–2394.

Doi, M. (1976) Stochastic theory of diffusion-controlled reactions, *J. Phys.* **A 9**, 1479–1495.

Felderhof, B.U. (1971) Spin relaxation of the Ising chain, *Rep. Math. Phys.* **1**, 215–234.

Glauber, R.J. (1963) Time-dependent statistics of the Ising model, *J. Math. Phys.* **4**, 294–307.

Grassberger, P. (1983) On the critical behavior of the general epidemic process and dynamical percolation, *Math. Biosci.* **63**, 157–172.

Grassberger, P. & de la Torre, A. (1979) Reggeon Field Theory (Schlögel's First Model) on a Lattice: Monte Carlo Calculations of Critical Behaviour, *Ann. Phys. – New York* **122**, 373–396.

Grassberger, P. & Scheunert, M. (1980) Fock-space methods for identical classical objects, *Fortschr. Phys.* **28**, 547–578.

Hiesmayr, B.C., Koniorczyk, M. & Narnhofer, H. (2006) Maximizing nearest-neighbor entanglement in finitely correlated qubit chains, *Phys. Rev.* **A 73**, 032310(11).

Hinrichsen, H. (2000) Nonequilibrium critical phenomena and transition into absorbing states, *arxiv: cond-mat/0001070v2* (also available in *Advances in Physics*).

Ising, E. (1925) Beitrag zur Theorie des Ferromagnetismus, *Z. Phys.* **31**, 253–258.

Janssen, H.K. (1981) On the nonequilibrium phase transition in reaction-diffusion systems with an absorbing stationary state, *Z. Phys.* **B 42**, 151–154.

Jensen, H.J. (1998) *Self-organized criticality, emergent complex behaviour in physical and biological systems* (Cambridge University Press, Cambridge).

Koniorczyk, M. & Janszky, J. (2001) Photon number conservation and photon interference, *arXiv: quant-ph/0110170v2*.

Lübeck, S. (2004) Universal scaling behavior of non-equilibrium phase transitions, *Int. J. Mod. Phys. B* **18**, 3977–4173 (also available at *arXiv: cond-mat/0501259*).

Ma, Shang-Keng (1976) Renormalization group by Monte Carlo methods, *Phys. Rev. Lett.* **37**, 461–464.

Marro, J. & Dickman, R. (1999) *Nonequilibrium phase transitions in lattice models* (Cambridge University Press, Cambridge).

Martins, J., Aguiar, M., Pinto, A. & Stollenwerk, N. (2009) On the series expansion of the spatial SIS evolution operator, *accepted for publication in J. Differ. Eq. Appl.*

Martins, J., Pinto, A. & Stollenwerk, N. (2009) A scaling analysis in the SIRI epidemiological model, *J. Biol. Dynam.* **3**, 479–496.

Park, S.C., Kim, D. & Park, J.M. (2000) Path-integral formulation of stochastic processes for exclusive particle systems, *Phys. Rev.* **E 62**, 7642–7645.

Park, S.C. & Park, J.M. (2005) Generating function, path integral representation, and equivalence for stochastic exclusive particle systems, *Phys. Rev.* **E 71**, 026113.

Peliti, L. (1985) Path integral approach to birth-death processes on a lattice, *J. Phys. – Paris* **46**, 1469–1483.

Stollenwerk, N. & Aguiar, M. (2008) The SIRI stochastic model with creation and annihilation operators, *available at arXiv:0806.4565v1.*

Stollenwerk, N. & Briggs, K.M. (2000) Master equation solution of a plant disease model, *Phys. Lett. A* **274**, 84–91.

Tomé, T. & de Oliveira, M.J. (2001) *Dinâmica estocástica e irreversibildade* (Editora da Universidade de São Paulo, São Paulo).

van Kampen, N.G. (1992) *Stochastic Processes in Physics and Chemistry* (North-Holland, Amsterdam).

Wijland, F. van. (2001) Field theory for reaction-diffusion processes with hard-core particles, *Phys. Rev.* **E 63**, 022101-1–4.

Yeomans, J.M. (1992). *Statistical Mechanics of Phase Transitions* (Oxford University Press, Oxford).

Zinn-Justin, J. (1989) *Quantum Field Theory and Critical Phenomena* (Oxford University Press, Oxford).

Chapter 6

Aron, J.L. & Schwartz, I.B. (1984) Seasonality and period-doubling bifurcations in an epidemic model, *J. Theor. Biol.* **110**, 665–679.

Anderson, R.M. & May, R. (1991) *Infectious Diseases in Humans* (Oxford University Press, Oxford).

Bolker, B.M. & Grenfell, B.T. (1993) Chaos and biological complexity in measles dynamics, *Proc. R. Soc. Lond.* **B 251**, 75–81.

Dietz, K. (1976) The incident of infectious diseases under the influence of seasonal fluctuations, *Lect. Notes Biomath.* **11**, 1–15.

Drepper, F.R., Engbert, R. & Stollenwerk, N. (1994) Nonlinear time series analysis of empirical population dynamics, *Ecol. Model.* **75/76**, 171–181.

Feistel, R. (1977) Betrachtung der Realisierung stochastischer Prozesse aus automatentheoretischer Sicht, *Wiss. Z. WPU Rostock* **26**, 663–670.

Gardiner, C.W. (1985) *Handbook of Stochastic Methods* (Springer, New York).

Gillespie, D.T. (1976) A general method for numerically simulating the stochastic time evolution of coupled chemical reactions, *J. Comput. Phys.* **22**, 403–434.

Gillespie, D.T. (1978) Monte Carlo simulation of random walks with residence time dependent transition probability rates, *J. Comput. Phys.* **28**, 395–407.

Grenfell, B.T. (1992) Chances and chaos in measles dynamics, *J. R. Stat. Soc.* **B 54**, 383–398.

Jansen, V.A.A. & Stollenwerk, N. (2005) Modeling measles outbreaks, in *Branching Processes: Variation, Growth, and Extinction of Populations*, eds. P. Haccou, P. Jagers & V. Vatutin (Cambridge University Press, Cambridge), 236–249.

Jansen, V.A.A., Stollenwerk, N., Jensen, H.J., Ramsey, M.E., Edmunds, W.J., & Rhodes, C.J. (2003) Measles outbreaks in a population with declining vaccine uptake, *Science* **301**, 804.

London, W.P. & Yorke, J.A. (1973) Recurrent outbreaks of measles, chickenpocks and mumps I, *Am. J. Epidemiol.* **98**, 453–468.

May, R.M. & Sugihara, G. (1990) Nonlinear forecasting as a way of distinguishing chaos measurement errors in time series, *Nature* **344**, 734–741.

Olsen, L.F. & Schaffer W.M. (1990) Chaos versus noisy periodicity: Alternative hypotheses for childhood epidemics, *Science* **249**, 499–504.

Rand, D.A. & Wilson, H.B. (1991) Chaotic stochasticity: A ubiquitous source of unpredictability in epidemics, *Proc. R. Soc. Lond.* **B 246**, 179–184.

Rhodes, C.J. & Anderson, R.M. (1996) Power laws governing epidemics in isolated populations, *Nature* **381**, 600–604.

Rhodes, C.J., Jensen, H.J. & Anderson, R.M. (1997) On the critical behaviour of simple epidemics, *Proc. R. Soc. Lond.* **B 264**, 1639–1649.

Schaffer, W.M. (1985) Order and chaos in ecological systems, *Ecology* **66**, 93–106.

Schaffer, W.M. & Kott, M. (1985) Nearly one dimensional dynamics in an epidemic, *J. Theor. Biol.* **112**, 403–427.

Schenzle, D. (1984) An age-structured model of pre- and post-vaccination measles transmission. *IMA J. Math. Appl. Med. Biol.* **1**, 169–191.

Schwartz, I.B. & Smith, H.L. (1983) Infinite subharmonic bifurcation in an SEIR epidemic model, *J. Math. Biol.* **B 18**, 233–253.

Stollenwerk, N. (1992) *Radial basis functions reconstruction from a time series of the measles cases in New York City* (Diplomarbeit, RWTH, Aachen).

Stollenwerk, N. (2001) Parameter estimation in nonlinear systems with dynamic noise. In *Integrative Systems Approaches to Natural and Social Sciences - System Science 2000*, eds. M. Matthies, H. Malchow & J. Kriz (Springer-Verlag, Berlin).

Stollenwerk, N. (2005) Self-organized criticality in human epidemiology. In *Modeling Cooperative Behavior in the Social Sciences*, eds. P.L. Garrido, J. Marro & M.A. Muñoz (American Institute of Physics AIP, New York), 191–193.

Yorke, J.A. & London, W.P. (1973) Recurrent outbreaks of measles, chickenpocks and mumps II, *Am. J. Epidemiol.* **98**, 469–482.

Chapter 7

Crow, F.C. & Kimura, M. (1970) *An Introduction to Population Genetics Theory* (Harper and Row, New York).

Goel, N.S. & Richter-Dyn, N. (1974) *Stochastic Models in Biology* (Academic Press, New York).

Karlin, S. & Taylor, H. (1975) *A First Course in Stochastic Processes* (Academic Press, New York).

Moran, P.A.P. (1958) A General Theory of the Distribution of Gene Frequencies, I. Overlapping Generations, *Proc. R. Soc. Lond. Series B, Biol. Sci.* **149**, 102–112.

Usmani, R. (1994) Inversion of a tridiagonal Jacobi matrix, *Linear Algebra Appl.* **212/213**, 413–414.

Chapter 8

Cartwright, K. (1995). *Meningococcal Disease* (John Wiley & Sons, Chichester).

Coen, P.G., Cartwright, K. & Stuart, J. (2000) Mathematical modelling of infection and disease due to *Neisseria meningitidis* and *Neisseria lactamica*, *Int. J. Epidemiol.* **29**, 180–188.

Guinea, F., Jansen, V.A.A. & Stollenwerk, N. (2005) Statistics of infections with diversity in the pathogenicity, *Biophys. Chem.* **115**, 181–185.

Maiden, M.C.J. (2000) High-throughput sequencing in the population analysis of bacterial pathogens of humans, *Int. J. Med. Microbiol.* **290**, 183–190.

Parkhill, J., Achtman, M., James, K.D., Bentley, S.D., Churcher, C. & Klee, S.R. (2000) Complete DNA sequence of a serogroup A strain of Neisseria meningitidis Z2491, *Nature* **404**, 502–506.

Stollenwerk, N. & Jansen, V.A.A. (2003a) Meningitis, pathogenicity near criticality: the epidemiology of meningococcal disease as a model for accidental pathogens, *J. Theor. Biol.* **222**, 347–359.

Stollenwerk, N. & Jansen, V.A.A. (2003b) Evolution towards criticality in an epidemiological model for meningococcal disease, *Phys. Lett. A* **317**, 87–96.

Stollenwerk, N., Maiden, M.C.J. & Jansen, V.A.A. (2004) Diversity in pathogenicity can cause outbreaks of menigococcal disease, *Proc. Natl. Acad. Sci. USA* **101**, 10229–10234.

Warden, M. (2001) *Universality: The Underlying Theory Behind Life, the Universe and Everything* (Macmillan, London).

Appendices

Abramowitz, M. & Stegun, I.A. (1972) *Handbook of Mathematical Functions* (Dover Publications, New York).

De Los Rios, P. (2001) Power law size distribution of supercritical random trees, *Europhys. Lett.* **56**, 898–903.

Green, P.J. (2001) A primer on Markov Chain Monte Carlo. In: *Complex Stochastic Systems*, eds. O.E. Barndorff-Nielsen, D.R. Cox & C. Klüppelberg (Chapman & Hall, Boca Raton), 1–62.

Grossmann, S. & Thomae, S. (1977) Invariant Distributions and Stationary Correlation Functions of One-Dimensional Discrete Processes, *Z Naturforsch* **32 a**, 1353–63.

Harris, T.E. (1989) *The Theory of Branching Processes* (Dover Publications, New York).

Index